Adaptations of Desert Organisms

Edited by J.L. Cloudsley-Thompson

Springer
Berlin
Heidelberg
New York
Barcelona
Budapest
Hong Kong
London
Milan
Paris
Santa Clara
Singapore
Tokyo

Volumes already published

Ecophysiology of the Camelidae and Desert Ruminants
By R.T. Wilson (1989)

Ecophysiology of Desert Arthropods and Reptiles
By J.L. Cloudsley-Thompson (1991)

Plant Nutrients in Desert Environments
By A. Day and K. Ludeke (1993)

Seed Germination in Desert Plants
By Y. Gutterman (1993)

Behavioural Adaptations of Desert Animals
By G. Costa (1995)

Invertebrates in Hot and Cold Arid Environments
By L. Sømme (1995)

Energetics of Desert Invertebrates
By H. Heatwole (1996)

Ecophysiology of Desert Birds
By G.L. Maclean (1996)

Plants of Desert Dunes
By A. Danin (1996)

Biotic Interactions in Arid Lands
By J.L. Cloudsley-Thompson (1996)

Structure-Function Relations of Warm Desert Plants
By A.C. Gibson (1996)

Physiological Ecology of North American Desert Plants
By S.D. Smith, R.K. Monson, and J.E. Anderson (1997)

Ecophysiology of Small Desert Mammals
By A.A. Degen (1997)

Homeostasis in Desert Reptiles
By S.D. Bradshaw (1997)

Ecophysiology of Amphibians Inhabiting Xeric Environments
By M.R. Warburg (1997)

In preparation

Avian Desert Predators
By W.E. Cook (1997)

Michael R. Warburg

Ecophysiology of Amphibians Inhabiting Xeric Environments

With 39 Figures and 18 Tables

 Springer

Prof. Dr. MICHAEL R. WARBURG
Department of Biology
Technion – Israel Institute of Technology
Haifa 32000
Israel

QL
669
.8
.W37
1997

ISSN 1430-9432
ISBN 3-540-59272-5 Springer-Verlag Berlin Heidelberg New York

© Springer-Verlag Berlin Heidelberg 1997
Printed in Germany

Cover design: Design & Production GmbH, Heidelberg

Typesetting: Best-set Typesetter Ltd., Hong Kong

SPIN: 10026896   31/3137/SPS – 5 4 3 2 1 0 – Printed on acid-free paper

This book is dedicated to the memory of my late mother, Ilse, who never discouraged me from bringing all kinds of curious animals into our home, and to my late sister, Hana, who so much wished to see the book I wrote but passed away too soon.

Preface

The main difficulty in this subject is to clarify the cardinal problems facing amphibians in seasonally xeric habitats. These problems are basically the same whether an amphibian inhabits xeric, semi-arid or arid habitats; the differences are solely a matter of magnitude. Thus, if we find that an amphibian must be well adapted and consequently capable of reducing its body water loss, the length of time it is exposed to such adverse conditions is the main distinction between these habitats. Likewise, it is the duration of time during which high temperatures prevail that varies. Therefore, we are dealing with two kinds of mechanisms: short-term and long-term survival. In order to be able to evaluate these adaptations, we need to look into the structure of various key organs such as skin, kidney, bladder, lung and ovary, among others. This knowledge will enable us to appreciate the ways the organs are involved in effecting these different kinds of adaptations.

Whereas these adaptations may concern the welfare of an individual animal, entirely different problems face the species that inhabits such habitats. To survive, a species must be capable of reproducing more of its own kind. Even if the individual can survive long periods of water shortage and lengthy exposure to high temperatures, this does not necessarily imply that the aquatic stages or young post-metamorphic juveniles can survive. In that case the animal must adapt in a different way by spreading out its reproductive efforts over a longer period. This, on the other hand, would involve different kinds of structural and ecological adaptations. Are amphibian species adapted to xeric environments long-lived, and if so, do they live longer than similar forms inhabiting less hostile habitats? Are they iteroparous (capable of reproducing more than once) or perhaps semelparous (producing one large cohort only)? Are they opportunistic breeders, prepared to seize any chance enabling them to breed? What kind of adaptations are required to be able to carry out such different reproductive patterns? These and many other kinds of questions were the ones I had in mind when setting out to write this book. Whether I succeeded in clarifying some of these points or raising several more instead will be judged by the reader. I personally feel profoundly rewarded by having had the good fortune to read so many interesting papers, while accumulating the kind of information needed here. I only regret that I could cite only a fraction of these most interesting studies; consequently, the decision concerning which papers should be discussed here and which not was a very difficult one for me.

As preparing this book required learning about various subjects some of which I understood less well than others, I am certain that the reader will notice the transitions. I apologise for any taxonomic errors, especially since I preferred using the original taxa name rather than the most current one (*Scaphiopus* rather than *Spea*, for example). I take comfort in the fact that the species' names change continuously, and thus one can never be fully up to date, as in all other fields. However, the basic picture is not likely to change because of that.

Haifa   MICHAEL R. WARBURG
December 1996

Acknowledgements

Much of the research for this book was done during my sabbatical leaves at the Universities of New Mexico (USA) and New South Wales (Australia). I am greatly indebted to Prof. Cliff Crawford and to Prof. Hal Heatwole for their kind hospitality in their labs at these respective institutes, and for having the opportunity to study in the libraries there. I wish also to thank Prof. Hal Cogger, formerly of the Australian Museum, Sydney, for providing the beautiful photographs of Australian frogs, Profs. Eduard Linsenmair and John Loveridge for the valuable prints of African frogs, Prof. Bill Duellman for enabling the representation of some South American amphibians, and Dr. Catherine Propper for providing Fig. 2.12. I am also indebted to Prof. John Cloudsley-Thompson for reviewing the original manuscript and for his most valuable suggestions.

Over the years I have benefited from the assistance of several students, some of whom are now my colleagues. Thus, I am indebted to Shoshi Goldenberg, Mira Rosenberg, Rakefet Sharon and Olga Gealekman for their enthusiastic collaboration. In particular, I enjoyed the support of Dina Lewinson and Gad Degani, former students and later colleagues. Mira Rosenberg, my assistant over the last 25 years, has devoted much of her life collaborating on the structural aspects of amphibian tissues. I am most indebted to her for making this partnership such a happy period. Finally, throughout my research over many years, I have enjoyed the encouragement of my family. First, my late father, Sigmunt, who introduced me to observing, handling and keeping animals, and to my late mother, Ilse, who tolerated all kinds of strange animals in our home. Later, I was actively helped by my wife, Hava, who joined me on many collecting trips in deserts of various continents, on hot days and sometimes rather cold, wet nights observing amphibian life. Later, my children joined in, each at his time. First, my son Ittai, who also prepared a beautiful school project on the salamander larva. Then my daughters, Sharon and Meirav, all of whom loved to go out on freezing cold, stormy nights to observe salamanders migrating to their breeding ponds, and who tried to identify them individually. I am very fortunate indeed to have had such support from all of them.

Contents

CHAPTER 1

Introduction

This book is a requiem to the amphibian class, which is on the verge of extinction. What we are witnessing is perhaps similar in a way to the extinction of the early reptiles, the dinosaurs. It is ironic and tragic that just as the whole amphibian class is virtually vanishing in front of our eyes without us knowing how to prevent this, several new books on amphibians should appear.

I do not intend this book to be a textbook on the structure, function or ecology of amphibians, nor is it a review of these subjects. There are several books already available, notably by Noble (1931), Moore (1964), Lofts (1976), Duellman and Trueb (1986) and Feder and Burggren (1992), as well as others. Rather it is an attempt to assemble some of the facts known about terrestrial amphibians, especially those inhabiting xeric, semi-arid and arid habitats, and to analyse them. Only on some occasions are other amphibians inhabiting mesic habitats discussed here, mostly as explanations or illustrations. It was not at all easy to decide which species should be included here and which not. One must keep in mind that the majority of studies were conducted in the more mesic parts of Europe and the USA. Only in comparatively recent years were such studies extended to other, more exotic species inhabiting xeric habitats in other parts of the world.

My own interest in amphibians extends far back to my childhood, when I used to go on long walks with my late father who was a real naturalist at heart, although not by profession. We collected amphibians (and reptiles) on Mt. Carmel and the Galil Mountains near Haifa, Israel, and always kept some for observation at our home. The first urodele larva I found in my childhood was during a long hike in a remote creek in the Galil Mountains. I placed it in my only water canteen, which meant that I had nothing to drink for the rest of this warm spring hike. It was later identified by my father as a *Salamandra* larva, which was at that time surprising since this habitat was most unlike any mesic habitat known to my father as typical for salamanders in Europe. Since that time, I have never lost interest in amphibians and have attempted to study them from different angles and as frequently as possible. Thanks to Prof. Dilon Ripley who enabled me to curate the herpetological collection at the Peabody Museum of Yale University and supported a collecting trip to the southwestern USA, I gained my first introduction to the North American fauna.

Although I have kept live amphibians (and reptiles) for many years since then, my scientific interest in them was aroused after reading Prof. Bert Main's review paper on the Australian amphibians (Main et al. 1959). At that time I planned my trip to Australia for post-doctoral studies. Later, on my return home, Prof. Hans Heller visited the Fisheries Department in Haifa, where I was temporarily employed, and introduced me to the world of neurohypophysial hormones and their influence on amphibians. His writings, as well as Peter Bentley's early papers, were undoubtedly influential on my thinking and subsequently on my own research.

CHAPTER 2

Xeric Habitats and Their Amphibian Inhabitants

2.1
The Habitats

The arid region includes areas in which the average annual rainfall does not exceed 100 mm, whereas in the semi-arid region the average annual precipitation ranges between 100 and 250 mm. Under xeric habitats we include areas where the annual rainfall can be higher provided it is restricted to a certain short period of the year; during the rest of the year, there is no precipitation nor any surface water. Temperatures during the summer are generally high in both arid and semi-arid habitats.

2.1.1
The Xeric Habitat

Such habitats are widespread and found in most regions. Thus, they can be found where the annual rainfall exceeds 250 mm, but long hot periods cause the ground to dry out completely. These habitats are normally devoid of any aquatic bodies, so that the amphibians have to aestivate in suitable microhabitats. Some amphibians belonging to the myobatrachids, leptodactylids and bufonids can be found living under these extreme xeric conditions.

These amphibians are largely anurans, totalling over 250 species, with bufonids, leptodactylids and myobatrachids comprising the bulk with about 50 species each, followed by hylids and hyperoliids with over 30 species each. There are also a few urodelan species (20). The leptodactylids and rhacophorids are confined to the southern hemisphere, whereas the pelobatids are confined to the northern hemisphere. The bufonids, hylids and ranids have a cosmopolitan distribution. In general, the rhacophorids, hylids and ranids are found in or near water, whereas the bufonids and pelobatids can be found at distances far from water. Some representatives of amphibians inhabiting xeric environments are seen in Figs. 2.1–2.12.

2.1.2
The Semi-arid and Arid Habitats

Semi-arid and arid habitats are found in Africa, Asia, the Americas and Australia. Many of them are inhabited by amphibians. Most of the Australian and

Fig. 2.1. Some Asian and African amphibians from xeric habitats: **a** *Triturus vittatus* terrestrial phase; **b** *Pelobates syriacus*; **c** three *Triturus vittatus* under one stone; **d** *Bufo viridis*; **e** *Bufo regularis*. (Courtesy of Prof. J.L. Cloudsley-Thompson)

Figs. 2.2.–2.6. Some Australian frogs inhabiting xeric habitats. (Courtesy of Prof. Hal Cogger)
Figs. 2.2. *Heleioporus psammophyllus* (*top*), *H. eyrei* (*middle*) and *Neobatrachus centralis* (*bottom*)

Fig. 2.3. *Notaden bennetti* (*top*), *N. nichollsii* (*bottom*)

Fig. 2.4. *Limnodynastes ornatus* (*top*), *L. peroni* (*middle*) and *Uperolia rugosa* (*bottom*)

Fig. 2.6. *Litoria caerulea* (*top*), *L. rubella* (*bottom*)

Fig. 2.5. *Cyclorana mainii* (*top*), *C. platycephalus* (*bottom*)

Figs. 2.7–2.10. Some African frogs
Fig. 2.7. **a,b** *Hyperolius viridiflavus nitidulus* aestivating; brilliant white at high temperature; **c** during wet season; **d** *Hyperolius nasutus*. (Courtesy of Prof. K.E. Linsenmair)

Fig. 2.8. *Hemisus marmoratus* (*top*) in amplexus. (Courtesy of Prof. K.E. Linsenmair)

Fig. 2.9. *Chiromantis xerampelina* (*middle*); *Hyperolius horstockii* (*bottom*). (Courtesy of Prof. J. Loveridge)

Fig. 2.10. *Pyxicephalus adspersus* (in cocoon) (*top*); *Brevipes rosei* (*middle*); *Leptopelis bocagei* (in cocoon) (*bottom*). (Courtesy of Prof. J. Loveridge)

Fig. 2.11. Some South American frogs. *Lepidobatrachus llanensis* (*top*); *Odontophrynus occidentalis* (*middle*); *Pleurodema bufonina* (*bottom*). (Courtesy of Prof. W. Duellman)

Fig. 2.12. *Scaphiopus couchi; above* in amplexus. (Courtesy of Dr. Catherine Propper)

American amphibian species studied here belong to this category. There are considerably fewer amphibian species (perhaps 50), inhabiting the true deserts where rainfall is less than 100 mm.

The amphibians inhabiting arid and semi-arid habitats belong largely to three families: Bufonidae Microhylidae and Leptodactylidae, although other families are represented also.

Some of the taxa occurring in either xeric, semi-arid and arid habitats are listed according to their locality in Table 2.1. This compilation is based on various sources. For the Asian amphibian fauna, the main sources were: Liu (1950), Zhao and Adler (1993), and personal communication with Dr. D.

Table 2.1. Amphibian species inhabiting xeric, semi-arid or arid habitats (numbers in brackets-source)

Asia

Anura

Bufonidae
Bufo viridis (1, 2, 3, 48)
B. danatensis (49)
B. raddei (3, 49, 76)
B. kavirensis (2)
B. arabicus (1, 51)
B. dhufarensis (1, 51)
B. surdus luristanicus (1)
B. tihamicus (51)
B. gargarizans (49)
B. galeatus (49)

Hylidae
Hyla ehrenbergi (1)
H. savigniyi (51, 55)

Microhylidae
Kaloula rugifera (75)
K. borealis (49)

Pelobatidae
Pelobates syriacus

Ranidae
Rana ridibunda (48, 49, 50)
R. levantina (55)
R. chensinensis (49)
R. nigromaculata (49)
R. amurensis (76)
R. tenggerensis (49)
R. asiatica (49)

Caudata

Salamandridae
Salamandra inframaculata (80)
Triturus vittatus (56, 78, 79)
Ranodon sibiricus (49)

Hynobiidae
Salamandrella keyserlingii (49, 77)

Africa

Bufonidae
Bufo bufo (7)
B. regularis (5, 6, 8, 65)
B. carens (5, 6)
B. rangeri (67)
B. ventralis hoeschi
B. vertebralis hoeschi (4, 9)
B. viridis (7, 8)
B. mauritanicus (7, 8)
B. dodsoni (8)
B. xeros (8)
B. brongersmai (7, 8)

B. pardalis (5)
B. gariepensis (5)
B. pusillus (6)
B. taitanus (6)
B. pentoni (14)

Hylidae
Hyla meridionalis (7)

Hemisidae
Hemisus marmoratus (5, 6, 14, 15, 64)

Hyperolidae
Afrixalus fornasinii (5, 12)
A. wittei (12)
A. weidholzi (12, 14)
A. brachycnemis (12, 64)
A. pygmaeus septentrionalis (12, 64)
A. flavovittatus (14)
Kassina senegalensis (5, 12, 14, 15, 69)
K. kuvagensis (12)
K. cassinoides (14)
Hylambates maculatus (5)
Leptopelis bocagei (5)
L. natalensis (5, 15)
L. oryi (12)
L. cynnamomeus (12)
L. viridis (12, 14)
L. argenteus (12)
L. bufonides (14)
Hyperolius argus
H. marmoratus taeniatus
H. nasutus (12, 14)
H. lamottei (14)
H. tuberilinguis (6)
H. paralelus albofasciatus (12)
H. viridiflavus taeniatus (12)
H. nitidulus (14)
H. pusillus (6, 12)
H. sheldricki (12)
H. parkeri (12)
H. balfouri (12)
H. quinquevittatus (12)

Microhylidae
Phrynomerus annectens (4, 9)
P. bifasciatus (4, 5, 15, 64)
P. microps (14)
Breviceps macrops
B. verrucosus (15)
B. adspersus (4, 5, 15)
B. gibbosus (5)
B. macrops (5)
B. mossambicus (5, 6)
B. montanus (5)
B. bifasciatus (6)

Table 2.1 *(Contd.)*

B. fowleri (6)
B. namaquensis (10)

Ranidae
Anhydrophryne rattrayei (15)
Artholeptella lightfooti (5)
Phrynobatrachus natalensis (6, 14)
P. francisci (14)
Ptychadena anchietae (6)
P. floweri (8)
P. mascareniensis (10)
P. submascariensis (14)
P. tournieri (14)
P. trinodis (14)
P. retropunctata (14)
Cacosternum namaquensis (10, 66)
C. bottgeri (15)
Rana perezi (7, 8)
R. occipitalis (8)
Tomopterna delalandii cyrtotis (4, 9, 14)
Hilderbandia ornata (4)
Pyxicephalus adspersus (4, 6, 10, 11, 14, 15)
P. delalandei (6, 15, 64)
T. natalensis (55)
T. marmorata (55)

Rhacophoridae
Chiromantis xerampelina (5, 6, 13)
C. petersii (64)
C. rufescens

Caudata

Salmandridae
Salamandra s. algira (7)

Australia

Hylidae
Cyclorana ausralis (9, 24)
C. brevipes (25, 74)
C. cryptotis (23, 24, 25)
C. cultripes (19, 24, 25, 68, 69)
C. longipes (24, 25)
C. maculosus (24, 25)
C. maini (24, 25, 74)
C. manya
C. novaehollandiae
C. platycephalus (18, 19, 24)
C. vagitus (24)
C. verrucosus (25, 74)
Lioria alboguttata (25)
L. caerulea (19, 24, 25)
L. cavernicola
L. electrica
L. gilleni
L. rubella (19, 24, 25)

L. splendida
L. peronii (16)

Leptodactylidae
Arenophryne rotunda (57, 58, 71, 72)
Pseudophryne bibroni (16)
P. occidentalis (16, 21, 22)
Glauertia douglasi (21)
G. russeli (22)
Heleioporus albopunctatus (19, 20)
H. australiacus (19, 20)
H. inornatus (19, 20)
H. eyrei (19, 20, 68)
H. psammophilus (19, 20)
H. barycragus (20)

Myobatrachidae
Linmodynastes convexiusculus (24)
L. dorsalis (25)
L. interioris
L. ornatus (24, 25)
L. salmini (16)
L. spenceri (16, 21, 24, 25, 68)
L. terraereginae
Neobatrachus albipes (70)
N. aquilonius (17, 24)
N. centralis (17, 22, 24, 53)
N. fulvus (17)
N. kunapalari (17)
N. pelobatoides (17)
N. pictus (53, 60)
N. sudelli
N. sutor (17, 22, 24, 69)
N. wilsmorei (17, 22)
Notaden benetti (16)
N. melanoscaphus (24)
N. nichollsi (16, 24)
Ranidella deserticola (24, 73)
Uperolia arenicola
U. aspera
U. borealis
U. capitulata
U. crassa
U. fusca
U. glandulosa
U. inundata
U. laevigata
U. lithomoda
U. littlejohni
U. marmorata
U. martini
U. micromeles (24)
U. mimula
U. minima
U. mjobergi

Table 2.1 (*Contd.*)

U. orientalis (24)
U. rugosa (16)
U. ruselli
U. talpa
U. trachyderma (24)
U. tyleri

North America

Bufonidae
Bufo woodhousei (32, 36)
B. cognatus (32, 36, 38)
B. mazatlanensis (32)
B. retiformis (32, 38)
B. alvarius (31, 32, 38)
B. microscaphus (32, 38)
B. debilis (38)
B. punctatus (29, 30, 32, 36)
B. compactilis (36)
B. americanus (33, 36)
B. hemiophrys
B. indsidor (36)
B. fowleri
B. boreas boreas (30, 39)
B.b. halophilus (39)
B. terrestris
B. vallipes
B. canorus (39)
B. speciosus (38)
B. exsul (39)

Hylidae
Hyla arenicolor (32, 38)
H. versicolor
H. regilla (39)
H. cadaverina (28, 38, 54)
H. californiae (30, 54)
H. eximia
Pternohyla fodiens (32, 38)
Acris creptians
Pseudacris nigrita

Microhylidae
Gastrophryne olivacea (38)

Leptodacylidae
Eleutherodactylus latrans (31)
Syrrhophus marnockii (40)

Scaphiopidae
Scaphiopus hammondi (32, 33, 38)
S. couchi (32, 33, 38)
S. bombifrons (38)
S. hurterii
S. intermontanus (38)
S. multiplicatus (61)

Ranidae
Rana terahumarae (26)

Rhinophrynidae
Rhinophrynus dorsalis (40)

Caudata

Ambytomatidae
Ambystoma tigrinum (34, 38)
A.t. nebulosum (27)
A.t. mavortium (27)
A. opacum (35)
A. talpoideum
A. dumerili
A. jeffersonianum (35)
A. maculatum (35)
A. mexicanum (35)
A. laterale (35)
A. texamum (35)

Plethodontidae
Batrachoseps attenuatus
B. major
B. aridus (37)
B. campi (37)

South America

Bufonidae
Bufo arenarum (41, 42, 44)
B. arunco
B. fernandezae (41)
B. spinulosus
B. spinulosum atacamensis (43)
B. major (41)
B. paracnemis (41, 59)
B. pygmaeus (41)
B. granulorus (62)
B. marinus (62)
Melanophryniscus stelzneri (41)

Hylidae
Hyla nasica (41)
H. pulchella (41)
H. trachythorax (41)
H. venulosa (41)
H. phrynoderma (41, 42)
H. nana (42)
H. lindneri (42)
H. rubra (62)
H. crepitans (62)
H. microcephala (62)
Phyllomedusa sauvagei (41, 46, 63)
P. iheringi
P. pailona
P. hypochondrialis (41, 42)

Table 2.1 (*Contd.*)

Leptodactylidae
Ceratophrys ornata (41, 46)
C. pierottii (41, 42, 63)
Lepidobatrachus llanensis (41, 46, 63)
L. asper (41, 42, 63)
L. laevis (41, 46)
L. salinicola (63)
Telmatobius reverberii (46)
T. solitarius (46)
T. somuncurensis (46)
Odontophrynus occidentalis (41, 46)
O. americanus (41, 46)
Leptodactylus ocellatus (41)
L. bufonius (41, 46, 63)
L. prognathus (41)
L. mystaceus (41, 42)
L. chaquensis (41, 46)
L. labyrinthicus (46)
L. laticeps (42)
L. sibilator (41)
L. gracilis (41)
L. mystacinus (41, 46)
L. bolivianus (62)

L. fuscus (62)
L. maerosternon (62)
L. fragilis (62)
L. wagneri (62)
L. anceps (63)
L. gualambensis (63)
Pleurodema cinerea (41)
P. nebulosa (41, 46)
P. tucumana (41, 46, 63)
P. marmorata (46)
P. guayapae (41)
P. borellii (63)
P. brachyops (62)
Physalaemus biligonigerus (41)
P. nattereri (46)
P. albonotatus (41, 46, 63)
P. cuvieri (42)
P. enesefae (62)
P. pastulosus (62)

Pipiidae
Pseudis paradoxus occidentalis (41, 46, 62)
Lysapsus limellus (41)

Sources: 1, Leviton et al. (1992); 2, Andren and Nilson (1979); 3, Orlova and Uteshev (1986); 4, Jurgens (1979); 5, Rose (1962); 6, Stewart (1967); 7, Mellado and Dakki (1988), Mellado and Mateo (1992); 8, Lambert (1984); 9, Channing (1976); 10, Van Dijk (1982); 11, Balinsky (1957); 12, Schiotz (1976); 13, Jennions et al. (1992); 14, Schiotz (1967); 15, Poynton (1964); 16, Cogger (1975, 1992); 17, Mahony and Roberts (1986); 18, Van Beurden (1982); 19, Main (1965); 20, Lee (1967); 21, Main (1968); 22, Main et al. (1959); 23, Tyler et al. (1982); 24, Tyler and Davies (1986); 25, Tyler et al. (1981); 26, Conant (1977); 27, Collins (1981); 28, McClanahan et al. (1994); 29, McClanahan et al. (1994); 30, Miller and Stebbins (1964); 31, Low (1976); 32, Blair (1975); 33, Bragg (1944, 1945); 34, Webb (1969); 35, Brandon (1989); 36, Bragg and Smith (1943); 37, Yanev and Wake (1981); 38, MacMahon (1985); 39, Karlstrom (1962); 40, Cochran (1961); 41, Blair (1975); 42, Cei (1959); 43, Cei (1955); 44, Kehr and Adema (1990); 45, Cei (1962); 46, Cei (1980); 47, Wager (1986); 48, Reed and Marx (1959); 49, Zhao and Adler (1993); 50, Sinsch and Eblenkamp (1994); 51, Showler (1995); 52, Tyler (1994); 53, Littlejohn (1966); 54, Ball and Jameson (1970); 55, Warburg (1971a); 56, Warburg (1971b); 57, Roberts (1984); 58, Roberts (1985); 59, Guix (1993); 60, Warburg (1965, 1967); 61, Woodward (1986); 62, Rivero-Blanco and Dixon (1979); 63, Gallardo (1979); 64, Bowker and Bowker (1979); 65, Winston (1955); 66, Balinsky (1962); 67, Passmore (1972); 68, Parker (1940); 69, Lindgren and Main (1961); 70, Roberts et al. (1991); 71, Roberts (1989); 72, Tyler et al. (1980); 73, Tyler (1994); 74, Tyler and Martin (1977); 75, Liu (1950); 76, Bobrov (1986); 77, Kuzmin (1984); 78, Kuzmin and Tarkhnishvili (1987); 79, Tarkhnishvili (1987); 80, Warburg (1986a,b, 1992a,b).

Tarkhnishvili of the University of Georgia, Tbilisi. For the African amphibian fauna, works by Rose (1962), Poynton (1964), Stewart (1967) and Wager (1986) were the main sources of information. I am well acquainted personally with the amphibian fauna from the semi-arid and arid regions in Australia, but I relied heavily on works by Gray (1845), Spencer (1896, 1928), Spencer and Gillen (1912), Parker (1940), Main (1965, 1968) and Cogger (1992). Much of the relevant information came from Mike Tyler of the University of Adelaide. The amphibian fauna of North America deserts is described in MacMahon (1985). Finally, the South American amphibians from arid habitats are described in Cei (1980) and Duellman (1979).

Although I have tried to include mostly the amphibian species occurring in xeric, semi-arid and arid habitats, some of them may be known by the reader to occur also in mesic habitats. On the other hand, other species that should have been listed may have been left out. About a third of the species are 'true' desert species. The main reason for compiling such a list is in order to get an idea of the extent of xeric adaptations in amphibians, their taxonomic affiliation, and their zoogeographical distribution. On the other hand, it may give the reader an idea of the extent to which some species have been studied, whereas most species are still awaiting study.

CHAPTER 3

Structural and Functional Adaptations of Key Organs

In this chapter, the structure and function of some organs that may play an important role in the survival of amphibians under xeric conditions will be discussed. These include the skin (epidermis and dermis), of great importance for protection, water and ion transport, and respiration, and other respiratory organs such as the gills and lungs. In order to understand water and ion balance, it is necessary to know the structure of the kidney and bladder. Finally, the female reproductive organs, especially the ovary, and oogenesis will be described in order to clarify the reproductive strategies of amphibians inhabiting xeric environments.

3.1
Structure and Function of the Integument

Under this term we include both the epidermis and dermis along with all structures related to them. There are a number of reviews on this subject (Lindemann and Voute 1976; Katz 1986; Toledo and Jared 1993b; Warburg et al. 1994a,b).

3.1.1
The Epidermal Structure of Amphibians Inhabiting Xeric Environments

The ultrastructure of amphibian epidermal cells has been described in at least 12 anurans, 4 urodeles and 2 caecilian species. Most of these studies, however, deal with just three genera: *Rana*, *Bufo* and *Triturus*. Only a few are concerned with the epidermal ultrastructure of amphibians inhabiting xeric environments (Warburg and Lewinson 1977; Rosenberg and Warburg 1991, 1992, 1993; Lewinson et al. 1982, 1984, 1987b). During metamorphosis the number of epidermal cell layers increases, the cell volume changes, and toward metamorphic climax (the tadpole with a tail stump) the outer cells flatten and keratinize to form the stratum corneum (Warburg et al. 1994a,b).

A number of cell types have been described in the amphibian epidermis (Warburg et al. 1994a,b). Among them are ones typical of larval stages, which ultimately disappear. These are presumably essential for life in an aquatic medium. Other cells make their first appearance towards the completion of the metamorphic climax and persist in the juvenile and adult stages.

The salamander (*Salamandra salamandra*) spends most of its life (over 95%) on land; only during its larval period, and for the adult female only for a few hours during breeding, is it again an aquatic animal. During this comparatively short period of its life, the larval epidermis needs to function as a respiratory organ in addition to its protective function as an integument. During the ontogenesis of the epidermis, several cell types make their appearance, and some of them are later replaced by others. This process is most likely related to the shift from aquatic to terrestrial life. The changes taking place in the physiology of the salamander during its ontogenesis until it metamorphoses from an aquatic larva into a terrestrial salamander are clearly reflected in the morphology of its epidermis.

The ontogenesis of the ventral epidermis and the integument covering the gill of *S. salamandra* has been described (Lewinson et al. 1983, 1987a; Warburg et al. 1994a,b). In newly hatched *S. salamandra* larvae, the epidermis is arranged in two layers and consists of four cell types. Towards metamorphic climax, some cell types are replaced by others (Warburg and Lewinson 1977).

The following account describes the changes that take place in the ventral epidermis of anuran and urodelan amphibians during metamorphosis and briefly considers the function and histochemistry of some integral cell components. A typical anuran, the fossorial pelobatid toad (*Pelobates syriacus*), and a xeric-inhabiting urodele (*S. salamandra infraimmaculata*) will be used as models. Other amphibians inhabiting xeric environments will be discussed as well.

3.1.1.1
The Epidermis Structure of the Anuran (Pelobates) Tadpole

In the legless *P. syriacus* tadpole, the epidermis consists of two cell layers (Fig. 3.1a). The outer (surface) layer contains four cell types, while the inner layer consists of one cell type only. The first surface cell is covered by cilia and disappears very soon at an early stage. These surface cells contain lipid droplets and electron-dense granules (Fig. 3.1a,c). The second cell type contains large, cup-like vesicles near its apical surface (Figs. 3.1d). The third type of cell is pear-shaped and bears a narrow surface of tuft-like microvilli. A fourth cell type has microvilli on its surface membrane and a large basal nucleus. It contains numerous mitochondria, especially at the upper part of the cell, and is therefore a mitochondria-rich cell (Fig. 3.1e). Directly above the basement membrane are the germinative cells, large, elongated cells rich in tonofilaments (Fig. 3.1a). In the two- and four-legged tadpole the epidermis consists of three cell layers, and individual cells decrease in volume. In addition to the previously described cell types, a flask-shaped cell has been recognized (Fig. 3.1e), situated under the replacement layer below the stratum corneum. Its surface is covered by microvilli, and it contains numerous mitochondria, an elaborate endoplasmic reticulum (ER) and Golgi complex, and a large, smooth, round nucleus located basally.

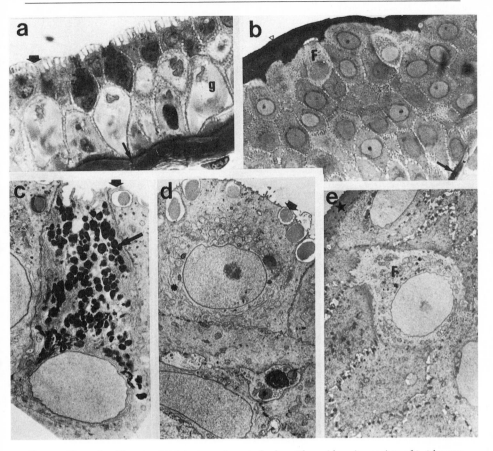

Fig. 3.1. Ventral epidermis of *Pelobates syriacus* tadpole. **a** The epidermis consists of 2–3 layers: the top layer is covered by microvilli (*arrowhead*) and contains large, dark periodic acid schiff (PAS)-positive cells. The bottom layer contains large germinative cells (*g*) situated on the basal membrane (*arrow*) (×100). **b** The epidermis of a post-metamorphic toad contains several layers situated on the basal membrane (*arrow*), a well-defined stratum corneum (*open arrowhead*) and under it large flask (*F*) cells (×100). **c** The PAS-positive cells containing a large number of granules (*arrow*). The face is marked by an *arrowhead* (×3000). **d** Vesicular, mucous-secreting (*arrowhead*) surface cell rich in rough endoplasmic reticulum (RER) and Golgi (×3000). **e** Adult epidermis with flask cell (*F*) (×3000)

At metamorphic climax, the epidermal layers flatten, and the surface layer begins to keratinize, forming a stratum corneum (Fig. 3.1b). Below this layer, a stratified replacement layer is present. Situated underneath is the stratum spinosum with its granular cells arranged in two or three cell layers. The large, pear-shaped granular surface cells have disappeared from the epidermis, and the number of flask cells has increased.

Dermal glands begin to form as nests in the epidermis, eventually penetrating into the dermis through the basement membrane during metamorphic climax. All the recognizable types of dermal glands can be found in the dermis at this stage (Fig. 3.5).

3.1.1.2
The Epidermis Structure of the Anuran (Pelobates), Post-Metamorphic, Terrestrial Stage

In the epidermis of the juvenile toadlet, both epidermal stratification and keratinization are completed. The germinative cells increase further in volume. These processes continue in the adult toad, where a further thickening of the epidermis takes place, and cells adhere closely to one another. No new epidermal cell appears. The only other change observed is in the relative numerical proportions of the cell.

3.1.1.3
Epidermal Thickness

In legless tadpoles, the epidermis is 18 μm thick. With continued development, the epidermal thickness in the two-legged tadpole grows to 65 μm, largely due to the increase in height of the surface and the basal germinative cells by as much as 76.8% (Warburg et al. 1994a,b). After metamorphosis, the epidermis becomes thinner simultaneously with cellular stratification. A juvenile toadlet prevented experimentally from climbing onto land and forced to remain in water retains a much thicker epidermis than a comparable juvenile on land.

3.1.1.4
Other Anurans Inhabiting Xeric Environments

Four other genera of anurans inhabiting xeric environments described here (*Hyperolius, Pseudophryne, Bufo, Hyla*) show similar structures (Figs. 3.2–3.4). The stratum corneum is stratified and varie in thickness (Fig. 3.2b,c,f). It consists of several layers in *Hyperolius* (Figs. 3.2e,h; 3.3c,d). Flask cells are invariably found in the epidermis of the adult anurans (Figs. 3.2b,c,g, 3.4a,b). Two main types of dermal glands can be distinguished: serosal and mucosal (Figs. 3.2e,f, 3.5).

Fig. 3.2. Ventral epidermis of various amphibians. **a** Ventral epidermis of *Salamandra* larva showing the unique, large Leydig cells (*L*) situated between the basal membrane (*arrow*) and the surface (*arrowhead*) (×40). **b** Adult salamander showing organized stratum corneum (*asterisk*) and a replacement layer as well as a flask cell (*F*) (×40). **c** Adult *Hyla savignyi* showing a flask cell (*F*) (×100). **d** A two-legged tadpole of *Hyla*, whose surface is not yet organized into a stratum corneum (*arrowhead*). Large germinative cells are situated on the basal membrane (*arrow*) (×100). **e** Skin of a juvenile *Heleioporus albopunctatus* showing a sloughed-off stratum corneum and a large dermal gland (*G*) in the dermis (×400). **f** Skin of adult *Pseudophryne occidentalis* showing two types of dermal glands: a serous gland (*1*) and a mucous gland (*2*) (×400). **g** Epidermis of adult *Bufo viridis* showing a flask cell (*F*) under the stratum corneum (*asterisk*) (×400). **h** Epidermis of *Hyperolius viridiflavusus* showing a sloughed-off layer of the stratum corneum in addition to another such layer and a replacement layer (*asterisk*) (×400)

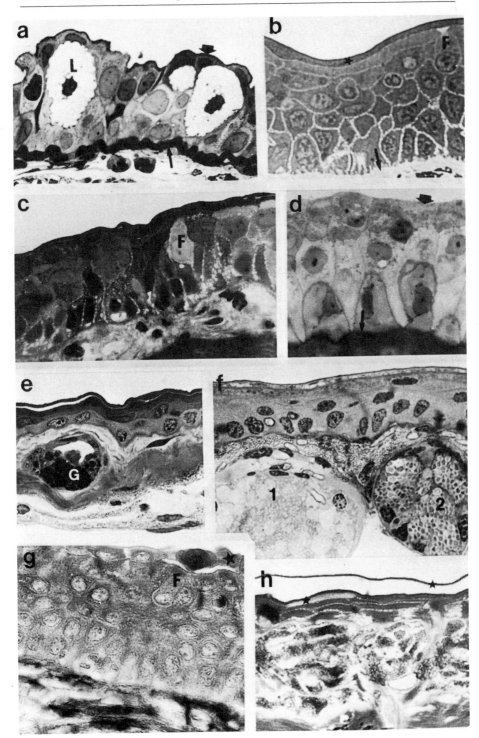

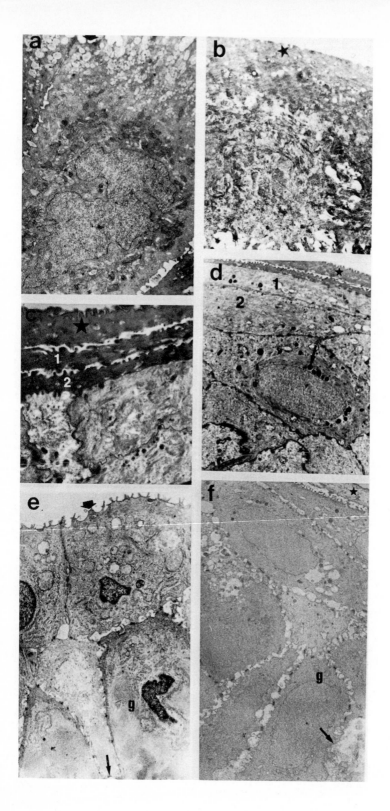

3.1.1.5
Cell Dynamics

The cells with granules at the surface layer are most abundant in legless tadpoles and decrease in number towards metamorphosis. These cells are still found in the epidermis of the four-legged tadpole (with a tail fin), but disappear from the epidermis towards the metamorphic climax. The surface granular cells are periodic acid-schiff (PAS)-positive (Fig. 3.1c). No such reaction is noticeable in any other cell throughout metamorphosis. The flask cells first appear in the epidermis of the four-legged tadpole following the completion of metamorphosis. They are greatly increased in number in the adult epidermis. A positive reaction to lipid stain (Oil-red-O) is evident in the epidermal surface cells. Similarly, a positive reaction to Sudan Black B can be observed in the surface cell's outer membrane.

3.1.1.6
The Epidermis of the Urodele Salamandra salamandra, Aquatic Stage

The epidermis of the newly born *S. s. infraimmaculata* larva contains two or three cell layers about 60 μm thick (Figs. 3.2a; 3.3a). These layers consist of five cell types. At the surface, there are "light" and 'dark' cells. Above the basement membrane are found the germinative dark cells. Scattered in between are numerous, large, vacuolated Leydig cells and among them a few mitochondria-rich cells (MRC) (Fig. 3.4d).

At the surface, the 'light cells' are large and vacuolated and have a large, homogeneous nucleus occupying almost the entire volume of the cell. The cytoplasm consists largely of microfibrils running in all directions which are sometimes arranged in bundles. Numerous, variably sized, membrane-bound vacuoles lie close to the nuclear membrane and can be found dispersed among the fibrils throughout the cytoplasm. The second type of cell found at the skin surface of the newly born larva is the "dark" cell. The bulk of the epidermis is comprised of dark cells which have a large, basally located nucleus. These dark cells are characterized by the presence of numerous club-shaped secretion vesicles which underlie the outer plasma membrane.

The epidermis of the fully grown larva consists of several layers containing cells that already resemble adult epidermis cells. When the larva is 4 weeks old, the main change in the epidermis is an increase in the number of basal cells. At

---

Fig. 3.3. Ventral epidermis of various amphibians. **a** Surface (*arrowhead*) of *Salamandra* larva ventral epidermis (×4000). **b** Ventral epidermis of adult salamander after formation of a stratum corneum (*asterisk*) (×9800). **c** Ventral epidermis of *Hyperolius* showing a well-formed stratum corneum (*asterisk*) in addition to two replacement layers (*1, 2*) (×6000). **d** Same as in **c**, in addition showing a granular cell containing granules (*arrow*) (×3000). **e** Ventral epidermis of legless *Hyla* tadpole showing two layers, the surface (*arrowhead*) cells and the large germinative cells (*g*) situated on the basal membrane (*arrow*) (×3000). **f** Adult *Hyla*; note the stratum corneum (*asterisk*), large germinative cells (*g*) situated on the basal membranes (*arrow*) (×2500)

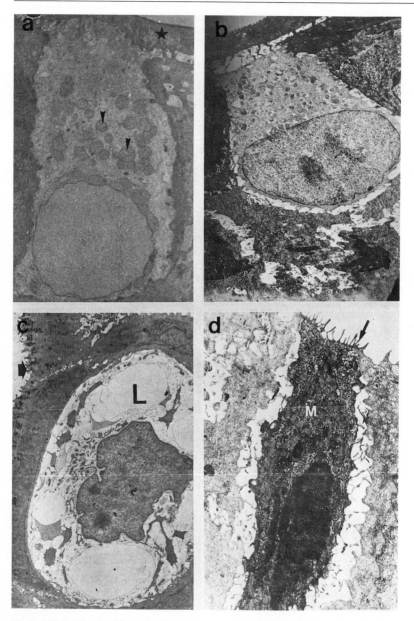

Fig. 3.4. Specialized epidermal cells. **a** A large flask cell under the stratum corneum (*asterisk*) of an adult *Hyla*. Note the numerous mitochondria in the flask cell (*arrowhead*) (×4000). **b** A large flask cell under the stratum corneum (*asterisk*) of an adult *Salamandra* (×3000). **c** Leydig cell (*L*) of *Salamandra* larva situated under the surface (*arrowhead*) (×2300). **d** Mitochondria-rich cell (*M*) at the face (*arrow*) of the ventral epidermis of a *Salamandra* larva (×4000)

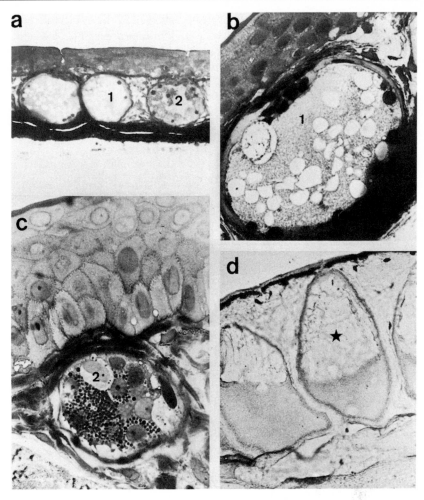

Fig. 3.5. Dermis of various anurans. **a** An adult *Pelobates*, note the two gland types: a serous gland (*1*) and a mucous gland (*2*) (×125). **b** A serous gland (*1*) of a juvenile *Pelobates* (×100). **c** A mucous gland in the dermis of an adult *Pelobates* (×100). **d** Large lipid glands (*asterisk*) in the dorsal skin of *Litoria caerulea* (×25)

this stage, a few light cells are found near the surface. Ultrastructurally, these resemble cells that have previously been described in the early developmental stage. Most of the cells on the surface are flat, dark cells. The permeability of the surface cells can be demonstrated by immersing the larvae in lanthanum nitrate solution. The tracer not only adheres to the flocular material inside the vacuoles, it also penetrates into the cytosol.

Indications of the presence of glycoconjugates on the apical border of the surface cells at this stage have been demonstrated by a positive reaction to PAS and Alcian Blue stains.

The first signs of keratinization can be observed in the epidermis of larvae. At the metamorphic climax, both dark and light cells are visible in the different layers of the epidermis, but at the surface only keratinized cells devoid of organelles can be found. True desmosomes can be seen for the first time in larvae at metamorphic climax.

3.1.1.7
Specialized Cell Types of the Urodelan (Salamandra) Larva

Two types of specialized cells characterize the urodelan larval epidermis: the Leydig cell and the mitochondria-rich cell (MRC). They will be briefly described here.

3.1.1.8
The Urodelan Leydig Cell

This cell is uniquely characteristic of urodelan larval life and is the most prominent cell in the larval epidermis. In the young larva the Leydig cell is often situated on the basal lamina and is surrounded by germinative cells (Figs. 3.2a, 3.4c). On occasion it is found in an intermediate position between the surface and the basal lamina, but it is never found at the epidermal surface.

The Leydig cell is oval in shape and contains a centrally located, lobed nucleus (Fig. 3.4c). All the main cellular organelles are located in the perinuclear cytoplasm, while the rest of the cell is filled with large vacuoles and different kinds of dense granules. The cell seems to be supported by a network of spiral filament bundles in hexagonal array: the so-called Langerhans net. In sections tangential to the epidermal surface, the peripheral location of the net can be appreciated. Only occasionally, when the section cuts through the periphery of the cell, does the hexagonal arrangement of the filament bundles become obvious. The Leydig cell hypertrophies with age, reaching dimensions of 30–50 μm in length by 20–35 μm in width.

3.1.1.9
The Larval Mitochondria-Rich Cell, MRC

The MRC is a narrow, elongated cell found scattered between the epidermal surface cells and reaching the surface in a limited area (Fig. 3.4d). Two different types of MRC can be distinguished in urodelan larvae. The type I MRC is a spindle-shaped cell, containing a large nucleus that occupies the major part of the cell. The remainder of the cell is filled with numerous mitochondria and small quantities of vesicles in between. Short processes extend from its surface (Fig. 3.4d). The type II MRC is usually darker in appearance and contains an elaborate membranous system in addition to the abundant mitochondria. This MRC type is decorated with a slender tuft of microvilli, and the membranous system reaches into the base of these microvilli. Both types of MRC show carbonic anhydrase activity.

3.1.1.10
The Epidermis of the Urodelan (S. salamandra), Terrestrial, Post-metamorphic Stages

Following completion of metamorphosis, the epidermis is characterized by cornification and stratification (Fig. 3.3b). The post-metamorphic and the adult's multilayered epidermis is organized in distinct strata. The germinative cells are separated from the dermis by the undulating basal membrane. These stratum germinativum cells are still proliferating, and mitotic figures can be observed in them. They provide the cells of the stratum spinosum.

The stratum spinosum cells are characterized by an abundance of cellular interdigitations, which reach across the intercellular spaces, and by numerous bundles of tonofilaments.

Between the stratum spinosum and the stratum corneum lies the stratum granulosum. The stratum granulosum consists of 1–2 cell layers. Its cells are less cuboid than the stratum spinosum cells. In addition to having many tonofilament bundles, they are also rich in electron-dense granules.

The outermost layer of the epidermis forms the stratum corneum. Typically, this consists of a cornified outer layer (Figs. 3.1b, 3.2c,e,h) and one or two replacement layers (Figs. 3.2b, 3.3c,d). The cells lack any organelles and contain only amorphous keratinized material. Vacuoles and the remnants of cell organelles can occasionally be seen.

3.1.1.11
The Flask Cell

While most other epidermal cells are organized in distinct strata parallel to the surface, a pear-shaped cell stands out in a vertical position. This cell is usually referred to as a flask cell (Figs. 3.2b, 3.4b). It has a typical subcorneal location and never reaches the surface. The apical region of the flask cells has secretion granules, mitochondria and bundles of filaments.

The flask cell contains a basally located, large and oval nucleus. Facing the apical part of the cell, the juxtanuclear region contains the Golgi apparatus and myelin figures, while the rest of the cytoplasm is densely packed with mitochondria.

Like the larval MRC, the adult's flask cell specializes in carbonic anhydrase activity (see Lewinson et al. 1982; Katz and Gabbay 1988).

3.1.1.12
Conclusions

Switching from aquatic to terrestrial habitats is reflected in the epidermal structure. These changes are seen not only in the increasing number of cell layers but also in their cellular composition. During the aquatic phase, the epidermis is still porous and plays a major role in gas exchange. When the animals emerge onto land, stratification and cornification take place. The integument must not only provide protection, but also allow for gases, water and solutes to pass selectively through it (Wittouck 1974). After metamorphosis the epidermis contains five to six cell layers arranged in three strata: stratum corneum, stratum spinosum and germinative layer. Dermal glands form as nests in the basal region of the epidermis and eventually penetrate through the basement membrane into the dermis.

3.1.1.13
Epidermal Cell Layers

The epidermal cell layers of amphibians increase in number during tadpole development, until completion of metamorphosis. There is no relationship between epidermal height and body size (weight) in *S. salamandra*. In legless larvae, there are two to three layers (Rosenberg et al. 1982). In juvenile *P. syriacus* there are three epidermal cell layers and in the adults, between three and four layers (Rosenberg and Warburg 1993). Similarly, the adult *Hyla arborea* (Bani et al. 1985; Rosenberg and Warburg 1995) and *S. salamandra* have three to four cell layers (Warburg and Lewinson 1977). Epidermal height is 33 µm in the juvenile *P. syriacus* toadlet and 56 µm in the adult, an increase of 70%. The height of the germinative cells (expressed as a percentage of the total epidermal height) is similar in both forms. Thus, the difference in total epidermal height must be due, in part at least, to an increase in individual epidermal cell volume in the adult.

During winter, there is evidence that the skin of *Bufo arenarum* becomes thicker (Lascano and Segura 1971).

The process of cell flattening and keratinization reaches its peak during the metamorphic period, and the stratum corneum is formed at its climax (Warburg and Lewinson 1977).

3.1.1.14
Epidermal Cell Diversity

Two cell types in the anuran epidermis are of special interest. The surface cell contains mucous vesicles beneath the surface membrane. In *Pelobates*, the vacuoles found in this cell appear to contain mucopolysaccharides (Bytinski-Salz 1976) and react slightly to PAS (Rosenberg and Warburg 1991). In *S. salamandra* the highest density of surface vesicles occurs after the larvae hatch (Lewinson et al. 1983). These vesicles contain PAS-positive material and are present in the larval epidermal surface cells (Lewinson et al. 1983), as well as in intra-uterine larvae (Greven 1980).

The second type is represented by the surface cells of *Pelobates* tadpoles, which contain PAS-positive granules. These cells disappear before metamorphosis concurrent with the appearance of the flask cells. Although the two types of cells are similar in being pear-shaped, only the latter are rich in mitochondria, whereas the surface cells are rich in PAS-positive granules (Rosenberg and Warburg 1991).

The MRC present in *P. syriacus* tadpoles are similar to one of the cell types previously described as type I MRC in *S. salamandra* (Lewinson et al. 1982).

A cell type that has received much attention in the past is the flask cell. About 10% of the epidermal cell population in the adult amphibian is made up of flask cells. This MRC is typically found only in adult anurans and urodeles. Although much research has been done on the flask cell, our understanding of its function is still limited. Flask cells contain a variety of enzymes (Zaccone et al. 1986), foremost among them carbonic anhydrase (CAH), found in *S. salamandra* (Lewinson et al. 1982). We were recently able to demonstrate ATPase activity in the basolateral membranes of this cell type (Rosenberg and Warburg 1991). Apparently, the formation of the flask cell may be related to the emergence of Amphibia on land (Warburg and Lewinson 1977; Lewinson et al. 1982).

3.1.1.15
The Leydig Cell of Larval Urodeles

The cell most typical of urodele larval skin is the peculiar Leydig cell. This extremely large cell is unique to urodeles (and some apodans). It occupies a large part of the epidermis and stretches from the basal lamina almost to the surface. In the intrauterine larva of *Salamandra*, the Leydig cells are still small (Pfitzner 1879, 1880) and not abundant. They are arranged in a single row under the apical membrane (Greven 1980). Leydig cells seem to decline in number after birth and change in structure, becoming larger towards meta-morphosis, after which they disappear entirely (Warburg and Lewinson 1977; Rosenberg et al. 1982). Greven (1980) followed the development of Leydig cells in *Salamandra* from birth to metamorphosis and discovered that they contain large vesicles and a Langerhans net (see Fig. 3.4c; Rosenberg et al. 1982). Our study indicates that the granules harbour PAS-positive glycoconjugates. Jarial (1989) found that, in the axolotl, mucous secretion is released into the intercel-lular spaces. However, the bulk of the Leydig cell is comprised of large vesicles that do not react positively with PAS reagent or Alcian Blue (AB), thus suggest-ing a possible role in water storage. Such a function is compatible with the hazardous events of larval life under unpredictable weather conditions. More-over, the fact that the Leydig cell does not open directly to the outer surface does not favour the hypothesis of a secretory role (of mucous secretion).

3.1.1.16
S. salamandra Larval MRC and the Flask Cells of the Adult

MRC are found in both the aquatic larva and the terrestrial post-metamorphic adult. Whereas the larval MRC (both epidermal and gill) open to the surface, the adult flask cell opens into the subcorneal space. Based on both ultrastructural and histochemical evidence, we assume that the adult flask cell is related to the larval MRC.

There are two types of MRC throughout larval life, which differ in shape, cytoplasmic density and length of apical villi (Lewinson et al. 1982). One type, which is distinguished by an elaborate reticulum, shows some morphological similarities with the "chloride cells" typical of fish epidermis (Lewinson et al. 1984).

Lodi (1971) demonstrated the presence of carbonic anhydrase (CAH) in the newt's flask cell. In *S. salamandra*, both types of larval MRC were found to be CAH-positive (Lewinson et al. 1982). CAH was found to be localized in the MRC of 1-week-old larvae. Its activity per cell increased with the development of the larva. The number of CAH-positive cells dropped towards metamorphosis (Lewinson et al. 1982, 1987b). The high content of CAH in the larval MRC facilitates proton and/or bicarbonate secretion, thereby indicating their role in gas exchange. As the adult flask cell contains CAH, it is likely that it plays a similar role here, too (Lewinson et al. 1982; Katz and Gabbay 1988). Aldosterone was found to induce CAH in the flask cells of frogs (Voute et al. 1975; Voute and Meier 1978).

We have not found any evidence for their role in sodium transport, as there is no indication of ATPase activity in the MRC. Na^+/K^+-ATPase activity was found only in the basolateral membrane of the pavement cells in the gill epithelium (Lewinson et al. 1987a), thus implicating the role of pavement cells in transport.

Zaccone et al. (1986) found oxyreductase transport enzymes in the flask cells of *Ambystoma tigrinum* and *A. laterale*. Moreover, they noted high levels of alkaline phosphatase (ATPase, K^+-p-NPPase) and carbonic anhydrase activity in the apical region of the flask cell.

The possibility that the adult flask cell is also active in separating the stratum corneum during the moulting cycle cannot be ruled out since secretory material spreads from the apical part of the flask cell throughout the subcorneal intercellular space.

3.1.2
The Dermis

3.1.2.1
The Anuran Calcified Dermal Layer

Between the stratum spongiosum that contains the dermal glands, elastic and muscle fibres, and the stratum compactum or solid layer composed of collagenous fibres, there is in most terrestrial anuran species (absent from urodeles) a layer described first by Eberth (1869) and Katschenko (1882). This layer was found to contain mucopolysaccharide deposits or "ground substance" (Elkan 1968). This ground substance, rich in calcium deposits, is highly hydrophilous, and 1 g can link with up to half his volume of water (Toledo and Jared 1993a,b).

The layer is better developed in the dorsal than in the ventral dermis (Elkan and Cooper 1980; Toledo and Jared 1993a). The fibrocytes related to this layer contain large quantities of calcium (Verhaagh and Greven 1982).

This layer is absent from most aquatic anurans. Although generally present in terrestrial anurans, a calcified layer is lacking in some typically terrestrial anurans, e.g. *Scaphiopus couchi* (juvenile), *Limnodynastes ornatus*, one of two specimens of *Bufo carens*, and several species of *Hyperolius* (Elkan 1968, 1976). In *Bufo arenarum* there is a marked increase in ground substance and mucopolysaccharides during the summer (Lascano and Segura 1971). Nothing is known about the ontogeny of this layer during anuran metamorphosis. Is it already formed towards metamorphic climax? There is inconclusive evidence showing dynamic changes in its thickness. Is it related to physiological events? The question of calcium metabolism will be discussed later (see Chap. 5).

3.1.2.2
Dermal Glands

Amphibian dermal glands first form in the epidermis and sink into the dermis upon metamorphosis (Fig. 3.5a). Thyroxine can affect this process (Yamashita and Iwasawa 1989).

Several dermal glands have been described in amphibians, the most common of which are the mucous and the serous glands (Figs. 3.2e,f, 3.5a–c). The mucous glands contain mucin which is mainly secreted in aquatic forms when exposed to the dry air (Lillywhite and Licht 1975). The serous (granular) glands secrete proteinaceous material and are essentially poison glands. The skin of *Hyla arborea* is extremely rich in glands (Grosse and Linnenbach 1989). In addition, glands secreting lipid material are found in a few arboreal anurans. Other specialized glands are known in particular rare cases.

In the larvae of *Ambystoma tigrinum*, the dermal glands are of two types: serous (granular) and mucous. The former secretes acid mucopolysaccharide substances in an apocrine manner, while the latter is smaller and more numerous, secreting in a merocrine manner (Le Quang Trong 1967). Both mucous and serous (granular) dermal glands have been shown to proliferate and differentiate when treated with testosterone (Norris et al. 1989).

Some of the granular dermal glands located in the tail of the adult *Ambystoma tigrinum* display a cyclic rhythm which may be taken as an indication of storing and metabolizing proteinaceous material rather than using fat (Williams and Larsen 1986). The number and size of the glands in the skin of *Bufo arenarum* decreases during the winter (Lascano and Segura 1971).

Blaylock et al. (1976) described alveolar dermal glands containing lipid material, mainly esters, in some South American phyllomedusine frogs. In *Phyllomedusa sauvagei*, for example, this secretion is effective in retarding the evaporation of water at temperatures up to 35 °C (McClanahan et al. 1978). These glands are larger than the mucous glands and more abundant, as can be seen in *Litoria caerulea* (Fig. 3.5d). They are about the same size as the serous glands but are only half as common in both dorsal and ventral skin. A cutaneous lipid layer was recently described in *Litoria fallax* and *L. peroni* (Amey and Grigg 1995). On the other hand, the skin lipid content was found to be higher in an aquatic ranid than a terrestrial bufonid (Schmid and Barden 1965). Apparently, these intradermal lipids found in a variety of anurans include a number of substances (esters, fatty acids, triglycerides, cholesterols and phospholipids), but do not play a role in the transport of water across the skin (Withers et al. 1984).

The male *Breviceps gibbosus* has peculiar holocrine glands on its ventral skin (in the female on the dorsal skin), which secrete glue during amplexus. This glue is believed to keep the pair together during the subterranean egg-laying period (Visser et al. 1982).

In some aquatic anurans, special, sexually dimorphic exocrine glands are dispersed throughout the skin (Thomas et al. 1993). These 'breeding glands' stain positively for neutral mucopolysaccharides and respond to testosterone stimulation similarly to the other dermal glands (Thomas and Licht 1993).

3.1.2.3
Chromatophores

Chromatophores are usually concentrated in the dorsal, gular and lateral parts of the skin (Drewes et al. 1977). In the skin of *Hyla arborea*, there are three types of chromatophores: melanophores, xanthophores and guanophores (= iridiophores) which, together, are responsible for the colour change (Nielsen and Bereiter-Hahn 1982; Grosse and Linnenbach 1989). Two types of iridiophores are known in *Pachymedusa danicolor* (Butman et al. 1979). The dispersion of melanosomes in the melanophores is under the control of MSH (melanophore-stimulating hormone; see Goldman and Mac Hadley 1969) and also of ACTH (adrenocorticotropin), whereas the iridiophores are especially sensitive to ACTH (Nielsen and Bereiter-Hahn 1982; see review by Bagnara and Fernandez 1993). *Hyperolius viridiflavus* and *Hyla arborea* are known to become lighter in colour at high ambient temperatures (Kobelt and Linsenmair 1986; Grosse and Linnenbach 1989). Ostrogen or testosterone alter chromatophore expression in *Hyperolius viridiflavus* (Richards 1982). Likewise, adrenaline or environmental stress cause an even lighter colour in *Hyla arborea* (Nielsen 1978b). This change is brought about by alteration in melanophore dispersal and in xanthophore and iridiophore shape (Nielsen 1978a; Kawaguti 1969).

In some rhacophorid frogs (e.g. *Chiromantis petersi*, *C. xerampelina*, *Hyperolius viridiflavus*), each chromatophore unit contains multiple iridiophores which are globular in shape. They stain with mucopolysaccharide dyes and are capable of reflecting light (Drewes et al. 1977; Kobelt and Linsenmair 1986). In *Hyperolius*, during the dry season, the number of iridiophores increased four to six times over their number during the wet season (Kobelt and Linsenmair 1986). Consequently, the colour of the frogs changes to bright white, thereby reflecting light. The colour of larval *Ambystoma tigrinum* changes according to their substrate background and water turbidity (Fernandez and Collins 1988). The skin of light-coloured larvae is composed of yellow xanthophores and silver-reflecting iridiophores containing four times more guanine than that of dark-coloured larvae (Fernandez 1988). In *Pachymedusa danicolor* and *Litoria caerulea*, the skin reflects near-infrared light, thereby preventing excessive heating (Schwalm et al. 1977).

3.1.3
Moulting

Moulting of the epidermis in *Bufo* takes place in two phases. First, the separation of the stratum corneum from its replacement layer, followed by the secretion of mucus and shedding of the skin (Jorgensen and Larsen 1964). Based on morphological changes we can distinguish between the intermoult, preparation, skin shedding and differentiation phases (Budtz and Larsen 1973).

Dent (1975) distinguishes between three consecutive processes during moulting: proliferation, differentiation (cornification of the keratinocytes) and ecdysis (casting off the stratum corneum). Skin shedding takes place following loosening of the desmosome connections and filling of the intercellular spaces with mucus (Parakkal and Matoltsy 1964; Larsen 1976).

Skin shedding is under hypophysial control and can be inhibited by extirpation of the pars distalis. This interferes with the preparation phase (Budtz 1977). Both ACTH and aldosterone can induce moulting in hypophysectomized toads, whereas thyroxine has no direct effect (Jorgensen and Larsen 1964; Jorgensen et al. 1965; Budtz 1977). Formation of a new stratum corneum is completed within 24 h (Budtz and Larsen 1973). The flask cells may play a part in moulting (Masoni and Garcia-Romeu 1979).

After spawning, many amphibians undergo frequent moulting (Larsen 1976). The moulting frequency also increases with rising temperature (Larsen 1976).

3.1.4
Cocoons

In aestivating amphibians, an epithelial cocoon is constructed of several layers of unshed, desquamated stratum corneum cell layers, forming an opaque envelope. These keratinized layers provide protection against dehydration. This has been described in several amphibian species (Table 3.1; Fig. 3.6): *Scaphiopus couchi* (Pelobatidae; Mayhew 1965), *Cyclorana australis, C. alboguttatus, C. platycephalus, Limnodynastes spenceri* and *Neobatrachus pictus* (Lee and Mercer 1967), *Lepidobatrachus llanensis* (Leptodactylidae; McClanahan et al. 1976), *Pternohyla fodiens* (Hylidae; Ruibal and Hillman 1981); *Pyxicephalus adspersus* and *Leptopelis bocagei* (Ranidae; Loveridge and Craye 1979; Loveridge and Withers 1981) and *Siren intermedia* and *S. lacertina* (Sirenidae; Reno et al. 1972; Etheridge 1990a,b). In the hylid *Pternohyla*

Fig. 3.6. Cocoon formation in *Neobatrachus* sp.

Table 3.1. Cocoon formation in amphibians from xeric habitats

Species	Reference
Pternohyla fodiens	Ruibal and Hillman (1981)
Smilisca baudinii	McDiarmid and Foster (1987)
Smilisca baudinii	McDiarrmid and Foster (1987)
Litoria alboguttata	Tyler et al. (1981)
Cyclorana platycephalus (compact layers	van Beurden (1984)
of keratinized cells)	Lee and Mercer (1967)
C. alboguttatus	Loveridge and Craye (1979)
	Lee and Mercer (1967)
C. australis	Loveridge and Craye (1979)
	Lee and Mercer (1967)
C. maini	Withers (1993)
Leptopelis bocagei	Loveridge and Craye (1979)
L. viridis	Loveridge (1976)
Lepidobatrachus llanensis (40–50 dead cell layers)	Cei (1980)
	McClanahan et al. (1976)
Ceratophrys ornata (keratinous cocoon)	Cei (1980)
	McClanahan et al. (1976)
Phrynomerus bifasciatus	Rose (1962)
Limnodynastes spenceri	Lee and Mercer (1967)
Neobatrachus pictus	Lee and Mercer (1967)
N. wilsmorei	Withers (1993)
N. kūnapalari	Withers (1993)
N. centralis	Withers (1993)
N. sutor	Withers (1993)
N. pelobatoides	Withers (1993)
Scaphiopus couchi (hard, dry, black material)	Mayhew (1962, 1965)
Pyxicephalus adspersus	Loveridge and Craye (1979)
Siren intermedia	Reno et al. 1972; Etheridge (1990a,b)
S. lacertina	Etherdige (1990a,b)

fodiens, multiple sloughs are interspersed by mucus secretion (Ruibal and Hillman 1981). No cocoon was ever seen in bufonids.

From observations on *Lepidobatrachus llanensis* it appears that the cocoon is formed at a rate of one layer per day over a period of a few weeks (Shoemaker 1988). This appears to be an outcome of a disturbance in ecdysis. It has been suggested that each of the three processes of moulting (proliferation, differentiation and ecdysis) are controlled by a different mechanism (Dent 1975). As aldosterone plays an important part in the detachment process or ecdysis during sloughing (Masoni and Garcia-Romeu 1979), the accumulation of unshed skin could be the result of a disturbance in its secretion. The metabolism of dormant cocooned anurans (Table 3.1) is depressed by 70–80% (Flanigan et al. 1991; Withers 1993) or 30% (Van Beurden 1980).

3.1.5
Transport of Ions Across the Skin

Sodium and water uptake through the skin (epidermis + dermis) evolved in freshwater ancestors of the amphibians (Kirschner 1983; Stiffler 1994). In North and South American hylids (e.g. *Hyla arenicolor*, *H. squirrella*, *H. femoralis* and *Agalychnis danicolor*) it is more pronounced in ventral skin than dorsal skin (Yorio and Bentley 1977). Water flux across the ventral skin is controlled by beta-adrenergic receptors and mediated by Na^+/K^+-ATPase (Nakashima and Kamishima 1990). There is adrenergic control over water uptake in *Bufo arenarum* (see Segura et al. 1982).

The efflux of Na^+ across the skin is greater in aquatic *Rana* spp. than in the terrestrial *Bufo marinus* (Bentley and Yorio 1976). Likewise, in *Ambystoma tigrinum* larvae and neotenous adults, sodium influx is greater than in the adults (Alvarado and Kirschner 1963; Alrarado 1979; Bentley and Baldwin 1980; Baldwin and Bentley 1982). When kept in a saline medium, larvae of *Ambystoma tigrinum* did not show any active Na^+ transport across the skin (Alvarado 1979).

In the South American amphibians *Bufo arunco*, *Leptodactylus ocellatus* and *Caudiverbera caudiverbera*, this ion movement is affected by mesotocin (Salibian et al. 1971; Salibian 1977). The permeability to sodium was affected by vasotocin. In *Ambystoma tigrinum* adults this was so, but not in the larva (Bentley and Baldwin 1980). Vasopressin affects the transport phenomena through the skin of *Bufo marinus* (Huja and Hong 1976).

Frog (*Rana*) skin is permeable to K^+ ions and excretes about 90% of the total (Frazier and Vanatta 1981). Urodele (*Ambystoma tigrinum*) skin is permeable in both directions to calcium ions (Baldwin and Bentley 1981a,b).

The exchange of ions across the skin in larval *Ambystoma tigrinum* is involved in the acid-base balance. The amphibious animal requires NaCl in order to compensate for respiratory acidosis (Stiffler et al. 1987; Rohbach and Stiffler 1987; Stiffler 1988).

When salt adapted, *Bufo viridis* shows a reduced rate of sodium uptake through the skin (Katz 1975; Rick et al. 1980). Similarly, when dehydrated, sodium loss from *B. regularis is* minimal (Vawda 1978).

3.1.6
Water Movement Across the Skin

Water moves up the skin of *Bufo boreas, B. woodhousei* and *B. hemiophrys* and along cutaneous channels (Lillywhite and Licht 1974). It remains to be demonstrated that the amounts of water moving over the skin and being absorbed by it can be of physiological significance. Apparently, water uptake by the skin of *Bufo marinus* follows reabsorption of dilute urine from the urinary bladder (Cirne et al. 1981). Permeability to water is unaffected by vasotocin in both larval and adult *Ambystoma tigrinum* (Bentley and Baldwin 1980). In *Hyla arborea japonica*, two types of water transport system through the ventral skin have been recognized: (1) normal permeability to water mediated through adrenergic receptors and (2) transport activated by dehydration and mediated by Na^+/K^+-ATPase activity (Nakashima and Kamishima 1990). During water absorption the outermost granular cells of the epidermis of *Hyla japonica* show irregular indentations on the plasma membrane, forming interdigitations with neighbouring cells. Lectin-binding patterns in the skin of *Ceratophrys ornata* are indications of glycoconjugates and associated with water movement through the skin (Faszewski and Kaltenbach 1995). In *Bufo arenarum*, water absorption through the permeable skin is related to the osmolarity of the external medium (Segura et al. 1984). Water can be absorbed from the soil through the skin in both *Bufo cognatus* and *Scaphiopus couchi* (Hillyard 1976a,b).

The pelvic skin of anurans contains abundant capillaries compared with other regions of the skin (Christensen 1974). High vascularity was found also in the skin of the inner side of the legs of *Hyperolius viridiflavus nitidulus* (Geise and Linsenmair 1986).

Bufo punctatus is capable of absorbing up to 70% of its total water uptake through the pelvic skin (McClanahan and Baldwin 1969). A somewhat lower absorption rate was demonstrated in *B. boreas* (Baldwin 1974). Water uptake is controlled by vasopressin and vasotocin (Christensen 1975). In *Neobatrachus pelobatoides* water movement through the pelvic skin was more than twice the normal rate through the rest of the skin and in *B. marinus*, more than three times (Bentley and Main 1972. *Bufo alvarius* is capable of rehydrating rapidly due to its extensive pelvic circulation (Roth 1973). On the other hand, *Scaphiopus couchi* has a very poorly developed pelvic circulation (Roth 1973). The pelvic region in bufonids shows abundant cutaneous musculature which may facilitate water absorption (Winokur and Hillyard 1992).

In *Scaphiopus couchi* and *S. hammondi*, water was taken up while condensing on the skin under naturally occurring conditions of temperature and humidity (Lasiewski and Bartholomew 1969).

There is a difference in the appearance of the surface of the ventral skin of specimens of *Ceratophrys ornata* from arid and temperate habitats in the

Chaco (Canziani and Cannata 1980). The toads from the arid habitats appear to have more granulated ventral skin, and the granules show a rich capillary network. It should be worthwhile studying these types of skin histologically.

3.1.7
Nitrogen Excretion Across the Skin

Ammonia is excreted through the skin of *Ambystoma tigrinum* larvae (Fanelli and Goldstein 1964; Dietz et al. 1967), adult *Necturus* sp., (Balinsky 1970), and the aquatic South American frog *Leptodactylus ocellatus* (Garcia-Romeu and Salibian 1968).

In *bufo viridis* there seems to be an urea transport mechanism across the skin (Katz et al. 1981).

3.1.8
Conclusions

The skin, both epidermis and dermis, plays a major role in protecting the amphibian from adverse environmental conditions due to water shortage and increased temperature. At the same time it facilitates respiration and carbon dioxide release. Finally, the skin functions as a major respiratory organ, even in adult amphibians with well-developed lungs. This is brought about by forming several cell layers that are composed of a variety of cell types. Under these layers calcified matter provides further defence. Finally, the dermal glands are the most significant sources of moisture and for thermoregulation. The direct relation between mucus secretion and temperature is not yet well understood.

There are a number of points cardinal to survival under xeric conditions that need further clarification. For example, does the transport of ions and water differ between cocooned and normal adults? Are there any variations during the different stages in the moulting cycle? We need to study further the function of the unique Leydig cell. Why is this cell type present in larval urodeles and not in adults? Moreover, the structural and functional dynamics of the dermal glands is relatively unknown. Finally, there is hardly any information available on the role of the skin in the excretion of nitrogen products.

3.2
Respiratory Organs and Respiration

3.2.1
Structure and Function of the Gills

The urodele larva has three pairs of gills which differ in length and number of branchial filaments (Bond 1960). The filaments are leaf-shaped and attached to a stem (Fig. 3.7a).

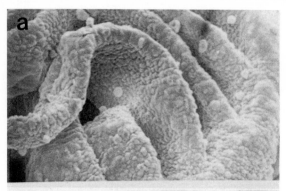

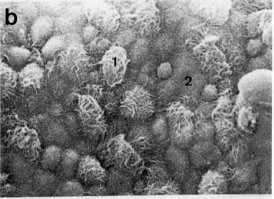

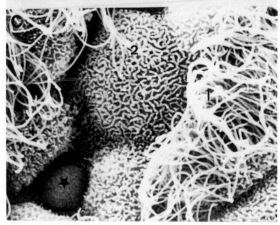

Fig. 3.7. Gills of *Salamandra* larvae. **a** The main gill lamela viewed by scanning electron microscopy (SEM; ×200). **b** Three cell types are noticeable: a ridged, pavement cell (*2*), a ciliary cell (*1*) and a mitochondria-rich cell (*asterisk*) (×1000). **c** A ridged pavement cell and small mitochondria-rich cell in between the pavement cells and the ciliary cells (×3500)

3.2.1.1
The Gill Integument of Salamandra Larvae

Each of the two gills consists of three main filaments of different lengths, each of which has two rows of secondary filaments (Fig. 3.7a). The length of the filaments and their numbers increase during early larval life (Szarski 1964). By scanning electron microscopy (SEM), three different kinds of cells can be observed (Fig. 3.7b). These three cell types are pavement cells, ciliary cells and MRC. The pavement cells are hexagonal and are covered by microridges. They are separated from each other by deep depressions. Generally, a few pavement cells surround a centrally located button-like protrusion which is probably the outer surface of an MRC. Finally, numerous ciliary cells, characterized by long and slender cilia, are dispersed between the pavement cells (Fig. 3.7c).

3.2.1.2
Specialized Cells in the Salamander Gill Integument

Two main kinds of MRC, both of which are pear-shaped, can be distinguished in the gill epidermis. The more abundant type is rich in mitochondria and filaments. The second type is characterized by a very elaborate tubulo-vesicular system which fills all the cytoplasmic spaces between the mitochondria. Both MRC types react positively to carbonic anhydrase (CAH), thereby suggesting their role in CO_2 elimination. This supports the theory that the internal gills of Devonian amphibians were the main site of carbon dioxide exchange.

The rate of oxygen consumption of gill-less *Ambystoma punctatum* larvae is the same as that of normal larvae (Boell et al. 1963; Heath 1976). Thus, the gills of aquatic larvae are not essential for respiration.

The *ciliary cell* is characterized by its flat nature and the abundance of cilia over its surface. In addition, it is covered by short cellular processes that are themselves covered by mucous tufts (Fig. 3.7).

3.2.2
Structure of the Lungs and Respiration

3.2.2.1
Structure

In the larval stages, the lungs consist of two simple sacs without septa (Fig. 3.8). The septa develop during amphibian ontogeny. Orgeig et al. (1994) distinguished several stages in the development of the lung and surfactant system in *Ambystoma tigrinum* from the fully aquatic larva (stage I) to the post-metamorph (stage VII). Simultaneously, the surface and volume of the lungs also increase considerably (Goniakowska-Witalinska 1980a, 1982; Czopek 1955; 1957, 1959, 1962). Thus, in larval *Triturus* the total thickness of the lung wall is 16–20 µm, whereas in juvenile individuals it is 21–46 µm, and in adults 132–138 µm (Goniakowska-Witalinska 1980b). The septation of the lungs is much more pronounced in anurans (Maina 1989). Although the structure of the lung does not change with growth, its volume does. Thus, in *Bufo viridis*, numerous, highly vascular septa point inwards, with an average of 729 meshes per mm^2 (Czopek and Czopek 1959). There is a definite correlation between

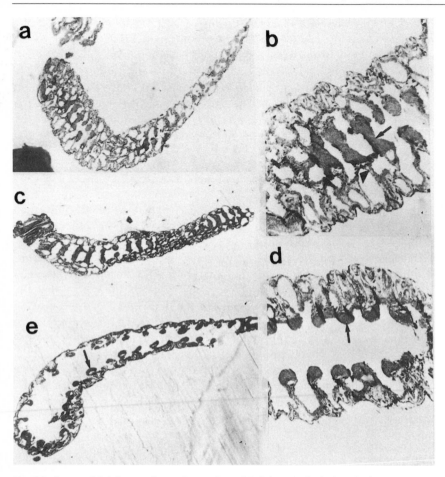

Fig. 3.8. Lungs of *Pelobates*: **a** lung of a two-legged *Pelobates* tadpole (×70), showing complete septation (**b**) (×200); **c** lung of a four-legged *Pelobates* tadpole (×50), showing formation of air space (**d**) (×200); **e** a mature lung of a metamorphosing *Pelobates* (×50)

respiratory surface area and metabolism in salamanders (Whitford and Hutchison 1967).

On the other hand, the number of capillary meshes per mm^2 diminishes with growth (Szarski 1964). The total length of the capillaries in the lungs of *Hyla arborea* is 46 m, three times larger than in *Bufo* (Czopek 1955). In *Scaphiopus couchi* the length of the respiratory capillaries in the lungs is considerably higher than in the skin. There is no such difference in *Scaphiopus hollbrooki* (Czopek et al. 1968). On the other hand, in the bufonids studied so far, the respiratory surface of the lung was 2–3 times larger than that of the skin (Bieniak and Watka 1962). This shows the significance of the lungs as the main respiratory organ.

There is a major difference between the oxygen requirements of urodeles and that of anurans. Urodeles have a relatively low oxygen requirement (Goniakowska-Witalinska 1982).

In the first stage of the larval *Salamandra*, the lungs are filled with fluid and contain lamellated bodies in their lumen (Pattle 1976; Pattle et al. 1977; Goniakowska-Witalinska 1982). In the second stage the lung fills with air, and the epithelial lining begins to differentiate. In the third stage, ciliated epithelium, pneumocytes and goblet cells, all of which contain osmiophilic lamellated bodies, vesicles and granules, make their appearance (Goniakowska-Witalinska 1978, 1980a,c,d,e, 1982; Maina 1989). The osmiophilic lamellar bodies are believed to be intracellular depots of surfactant (Pattle et al. 1977). Lungs of *Rana catesbeiana* tadpoles begin to function at metamorphosis. The surfactant phospholipid was mostly phosphatidylcholine (Oguchi et al. 1994). Ohkawa et al. (1989) found that at metamorphic climax digit-like processes intercalating with each other appear, coinciding with the appearance of lamellar bodies in the epithelium.

Endocrine-like cells (NE, neuroendocrine) are found in the lungs of various amphibians (Goniakowska-Witalinska 1980d, 1982; Cutz et al. 1986; Scheuermann et al. 1989; Goniakowska-Witalinska and Cutz 1990; Goniakowska-Witalinska et al. 1992). In *Bufo viridis* they are situated on special protrusions of the septa of the lungs (Goniakowska-Witalinska and Cutz 1990). In *Ambystoma tigrinum*, three to five of these cells are grouped to form a neuroepithelial body (NEB; Goniakowska-Witalinska et al. 1992). These cells are immunoreactive to serotonin, somatostatin and enkephalins (Cutz et al. 1986; Kikuchi et al. 1992).

3.2.2.2
Respiration

Amphibians have four main organs for gas exchange: gills, lungs, skin and buccopharyneal mucosa (Whitford and Hutchison 1965a). The lungless plethodontid salamanders utilize only cutaneous and buccopharyngeal respiration. Larval forms and some aquatic adults use gills. Thus, in *Necturus* about 40% of the respiration takes place through the gills, whereas in *Siren*, the gills do not play a major role, and most of the respiration is cutaneous (Shield and Bentley 1973). Similarly, in larvae and paedomorphic *Ambystoma tigrinum*, ligating the gills reduces oxygen consumption only slightly (Heath 1976). With the increase in thickness of the integument during metamorphosis, cutaneous respiration decreases by 60% (Ultsch 1973). In *Ambystoma tigrinum, A. talpoideum, Salamandra salamandra, Bufo regularis* and *B. marinus*, between 50 and 67% of the oxygen uptake and about 80% of carbon dioxide elimination (at 15 °C) are cutaneous (Whitford and Hutchison 1963, 1965a, 1966; Hughes 1966, 1967; Hutchison et al. 1968; Cloudsley-Thompson 1970; Shield and Bentley 1973; Packard 1976). When in water, *A. tigrinum* obtains less than 30% of its oxygen via the skin (Whitford and Sherman 1968). Likewise, when the permeability of water through the skin is reduced (in *Phyllomedusa sauvagei* and *Chiromantis xerampelina*), so is the cutaneous gas exchange (Stinner and Shoemaker 1987).

Since oxygen uptake through the skin is passive, its efficiency is dependent upon the number of capillaries and their proximity to the skin surface (Czopek 1962). The role of the skin in CO_2 excretion remains constant or increases after metamorphosis (Burggren and Wood 1981; Feder and Burggren 1985). This is supported by the finding that carbonic anhydrase levels in the terrestrial adult *A. tigrinum* are higher than in the aquatic larvae (Toews et al. 1978). CO_2 excretion through the skin is greatly reduced in dehydrated *Bufo marinus* (Boutilier et al. 1979a).

In salamanders, pulmonary oxygen uptake increases linearly with temperature (Whitford and Hutchison 1965b; Whitford 1973; Carey 1979a; Burggren and Wood 1981). Likewise, O_2 consumption is affected by temperature in *Hyla arenicolor* (Preest et al. 1992). The lungs of *Ceratophrys calcarata* respire 60% of the total oxygen uptake at 25 °C, and over 70% at 35 °C (Hutchison et al. 1968). There are differences between the genera and families. Thus, *Bufo viridis* has a higher O_2 consumption than *Rana ridibunda* at higher temperatures (27–36 °C, Degani and Meltzer 1988). Heat stress is accompanied by a drop in CO_2 elimination (Paulson and Hutchison 1987).

Buccopharyngeal respiration becomes increasingly important at higher temperatures in *Ambystoma maculatum* (Whitford and Hutchison 1963). On the other hand, CO_2 conductance through the skin does not change with temperature (Feder and Burggren 1985).

There is a diurnal cycle of oxygen consumption in *A. tigrinum* (Hutchison et al. 1977). This could well be related to the locomotory rhythm, as it is in *Bufo marinus* (Hutchison and Kohl 1971).

During aestivation, *Scaphiopus* spp. in burrows utilize between 15 and 25% of their normal active oxygen consumption (Seymour 1973a,b; Whitford and Meltzer 1976). This could largely be due to the reduction in cutaneous respiration through the impermeable cocoon. Similarly, in *Bufo marinus*, while the animals are in burrows, the role of the skin in gas exchange is reduced (Boutilier et al. 1979b).

With increase in body size, the skin becomes less efficient at gas exchange (Feder and Burggren 1985). This is especially the case in anurans with their high oxygen consumption rate, which is correlated with increased cardiac output, resulting in increased peripheral blood circulation and blood oxygen capacity (Hillman 1976; Sherman 1980a,b). The situation changes during dehydration when the loss of plasma volume is the main stress causing vasoconstriction, which leads to lower O_2 consumption (Hillman 1987). In dehydrated *Bufo viridis*, the resulting respiratory acidosis was found to be due to a drop in CO_2 elimination (Katz 1980).

3.3
The Endolymphatic System as Calcium Reserve

Although most of the calcium in amphibians is present in their bones, $CaCO_3$ is stored in the endolymphatic sacs located between the vertebrae (also called "paravertebral lime sacs"). This calcium store can be utilized later (Guardabassi 1960; Pilkington and Simkiss 1966; Robertson 1971a,b,c).

Amphibians have a well-developed subdermal lymphatic system consisting of lymphatic pulsating hearts, lymph channels and lymph sacs. The number of lymph hearts varies: in most adult anurans there is one pair of lymph hearts; in *Salamandra salamandra*, there are 15 pairs of lymph sacs: while in apodans there are over 200 (see references in Noble 1931 and Deyrup 1964). The lymph hearts are capable of rhythmic contraction. The terrestrial amphibians, in particular the xeric-inhabiting ones (*Heleiporus alboguttatus, Neobatrachus centralis, N. pictus, Notaden nichollsi*), have considerably smaller lymph sacs. Large-sized lymph sacs seem to be mainly associated with aquatic amphibians (Carter 1979). However, there are exceptions to this: *Litoria caerulea* and *Cyclorana australis* have extensive lymph sacs. On the other hand, the endolymphatic sac of *Hyla arborea japonica* is composed of many small chambers (Kawamata et al. 1987).

The endolymphatic sac of *Hyla arborea japonica* contains calcium ions (or pyroantimonate-precipitable calcium, see Kawamata 1990a). When $CaCl_2$ loading of that frog was done, $CaCO_3$ crystal formation was accelerated in the endolymphatic sac (Kawamata 1990b). Upon decalcification the ghosts of crystals remain in the lumina of the endolymphatic sac (Kawamata et al. 1987). Apparently, when dehydrated, anurans can mobilize the lymphatic fluid (Hillman et al. 1987). Calcitonin (CT), secreted by the ultimobranchial bodies, can block the resorption of calcium from its reservoir into the plasma (Robertson 1971a,b,c; Bentley 1984). The ultimobranchial bodies in *bufo viridis* show structural changes which are probably related to fluctuation in calcium levels and under the control of prolactin (Boschwitz 1967, 1977).

3.4
Kidney Structure and Function

3.4.1.1
Kidney Structure

The amphibian larval kidney or pronephros is osmoregulatory and eliminates waste in larval forms (Fox 1963). It disappears during metamorphic climax and is replaced by the mesonephros or adult kidney (Nussbaum 1886). In *Bufo viridis* the degeneration of the pronephros takes place comparatively early (Michael and Yacob 1974). There is some difference between anuran and urodelean kidneys (see details in Deyrup 1964). Nothing is known about the

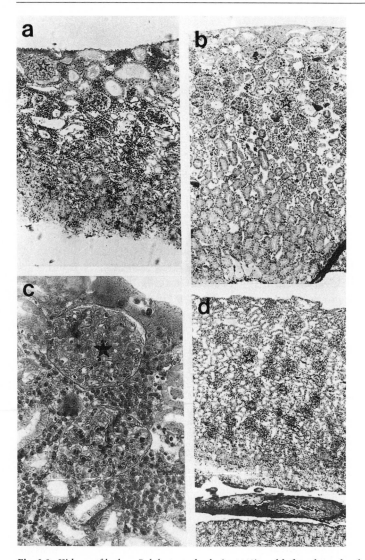

Fig. 3.9. Kidney of legless *Pelobates* tadpole (**a**, ×40) and **b** four-legged tadpole (×40) showing glomeruli (*open asterisk*). The post-metamorphic juvenile (**d**) has more glomeruli (×40; *asterisk*) and larger ones (*asterisk*; **c**, ×400)

ontogenesis of the kidney in more xeric species or in terrestrially breeding species.

The adult kidney contains nephrons consisting of a renal corpuscle or a glomerulus and a Bowmann capsule (Fig. 3.9). These are connected to a convoluted proximal tube and, through an intermediate segment, to a distal convoluted tube. This tube connects through an intermediate segment to the collecting duct (Pons et al. 1982). It is thought that the glomerulus evolved in freshwater vertebrates as a means to excrete large quantities of water. Its development is related to the availability of fresh water (Marshall and Smith 1930; for earlier reviews see also Deyrup 1964). The glomerulus reached its highest degree of development in the Dipnoi and Teleostomi (Hickman and Trump 1969).

There is some difference in the alkaline and acid phosphatase activity of the cells of the various segments, with a drop in activity during (winter) hibernation.

In the xeric-inhabiting genera *Chiromantis* and *Hyperolius*, the kidneys are rather similar anatomically speaking (Bhaduri and Basu 1957). The kidney structure of some xeric-inhabiting Australian anurans (*Heleioporus pictus, Notaden benetti, Cyclorana alboguttatus*) has been described by Sweet (1907). There appears to be a marked reduction in the number and size of the glomeruli of some xeric-inhabiting species as compared with other anurans (*Cyclorana alboguttatus* and *C. platycephalus*; see Dawson 1948, 1951). Moreover, the number and size of the cell nuclei in the kidney tubules increase. In addition, there is a lack of cilia in the neck and intermediate segment in *Cyclorana* (Sweet 1907).

3.4.1.2
Kidney Function

Urine production by the kidney in *Bufo arenarum* is under the control of central and peripheral adrenergic nervous components (Petriella et al. 1989).

The kidney of some terrestrial anurans (*Phylomedusa sauvagei, Bufo paracnemis* and *Ceratophis ornata*) is capable of passive urea reabsorption, thus facilitating hypotonicity of the urinary reserve in the bladder (Carlisky et al. 1968).

The kidney of *Ambystoma* is capable of releasing renin. There are higher levels of renal renin in the early developmental stages of several Australian anurans (Taylor et al. 1982). However, no relationship could be found between renin content and lifestyle (Taylor et al. 1982).

In *Ambystoma tigrinum* larvae, renal function during osmotic load occurs largely through decreasing glomerular filtration rate (GFR; Kirschner et al. 1971; Stiffler and Alvarado 1974). Similarly, dehydration leads to reduced urine production due to a 67% decrease in GFR (Tufts and Toews 1986). On the other hand, Pruett et al. (1991), analysing available data on GFR, concluded that terrestrial anurans have a greater GFR and urine flow than even the aquatic urodeles. Thus, in *Ambystoma* it was 0.64–1.04, whereas in *Phyllomedusa* it was 1.66, and in some bufonids between 3.79 and 12.22.

3.4.2
Endocrine Control of Excretion

The hypophysis is responsible for controlling much of the endocrine activity related to water and ion balance (Fig. 3.10).

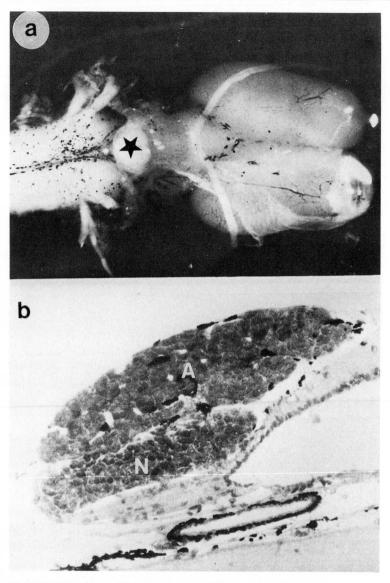

Fig. 3.10. Hypophysis of *Salamandra* (*asterisk, top*, ×6) and section through the hypophysis of *Pelobates* (*bottom, N* neuro, *A* adeno, ×100)

Glomerular antidiuresis is noticeable in response to treatment with arginine vasotocin (AVT). It reduced both urine flow and GFR by 30% in larval *A. tigrinum* (Stiffler et al. 1982). In *Necturus*, AVT changes the permeability of the distal tubule (Garland et al. 1975). However, Pang et al. (1983) failed to show this.

Mesotocin (MT) is a diuretic agent in *Rana*, *Bufo* and *Ambystoma* spp.,

Fig. 3.11. Hypophysial cells. **a** Large follicle composed of prolactin cells (*star*, Azan; ×100); **b** transmission EM (TEM), note the secretory granules (*asterisk*) of one type of cell (×4000); **c** TEM, note the large secretory granules (*star*) smaller of a second cell type (×4000)

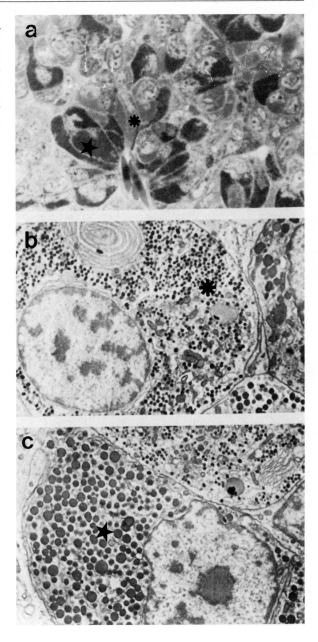

causing an increase in urine flow and GFR (Pang and Sawyer 1978; Stiffler 1981; Stiffler et al. 1982, 1984).

Prolactin (PRL) secreted by the prolactin cells in the hypohysis (Fig. 3.11) has an osmoregulatory effect on amphibians, affecting the proximal tubules (Hirano 1977; Gona 1982).

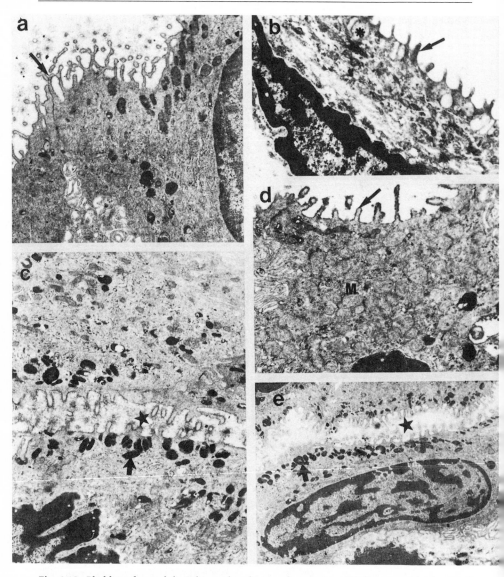

Fig. 3.12. Bladder of an adult *Salamandra* showing long microvilli (*long arrow*), elongated granular cells containing large granules (*short arrow*), numerous mitrochondria (*M*) large intercellular spaces in between cells (*asterisk*). **a**, ×4557; **b**, ×17670; **c**, ×12741; **d**, ×8835, **e**, ×4929)

Parathyroid hormone (PTH) has a wide range of effects on renal function and stimulates the conservation of calcium in vertebrates (Braun and Dantzler 1987).

Aldosterone has been found in amphibian plasma. It occurs at much higher titers in *Rana* than in either larval or adult *Ambystoma* (Stiffler et al. 1986), larvae having a concentration twice as high as adults. Aldosterone stimulates

Na^+ reabsorption and causes K^+ secretion in *Ambystoma tigrinum* larvae (Stiffler et al. 1986).

3.4.3
The Urinary Bladder

The amphibian urinary bladder is a bilobed sac, a derivative of the cloaca. Its size does not vary significantly between aquatic and terrestrial anurans. Thus, both the aquatic *Calyptocephalella caudiverbera* and the terrestrial *Bufo arunco* have a urinary bladder capacity of 8–11% of their total body volume (Espina and Rojas 1972). In hydrated *Pyxicephalus* the bladder contained 14.1% of the body mass (Loveridge and Withers 1981). As the urinary bladder is highly distensible, it can hold fluid up to 50% of the total body weight (Bentley 1966b). In that way it is an efficient water storage organ which can be utilized by xeric-inhabiting terrestrial amphibians under conditions of drought (Ruibal 1962; Bentley 1966b; Middler et al. 1968).

The urinary bladder is simple in structure and is composed of a single or double layer of epithelial cells (Fig. 3.12). In anurans, three main cell types are known: (1) granular (squamous) cells facing the surface (lumen); their apical sides consist of many microvilli which are covered on their luminal side by a fuzzy coat containing mucopolysaccharides; (2) mucous cells which are found at different frequencies in different anuran species; and (3) mitochondria-rich cells (MRC). In urodeles, a fourth cell type, the basal cell, can occasionally be seen. It is of interest to note that ciliated cells are apparently absent from the amphibian bladder, although they are common in the reptilian bladder (Bolton and Beuchat 1991).

In *Bufo marinus*, the urinary bladder cells (MRC, granular and basal) contain a large number of vesicles which seem to be involved in water movement across the bladder (Davis et al. 1982). Transport of sodium takes place from the mucosal (inner) surface to the serosal (outer) surface of the bladder. As a result, water can be reabsorbed from the bladder into the body fluid (Bentley 1966b). The exact mechanism involved has not yet been determined, nor is it known how water reabsorption is triggered.

The acidity of the mucous varies in both mucous and granular cells. A lower acidity of mucous secretion was noticed in *Salamandra salamandra* (de Piceis Polver et al. 1981). Furthermore, this was associated with a high degree of ATPase activity in the MRC, which also exhibited a high reductase activity. Both catecholamines, epinephrine and norepinephrine, caused a significant increase in H^+ and NH_4^+ excretion in the *Bufo marinus* bladder (Frazier 1983).

Antidiuretic hormones (AVT, AVP) as well as angiotensin II, aldosterone and, to some extent, the thyroid hormones cause increased permeability of the urinary bladder epithelium and the transfer of sodium and water across its cell walls in the anurans *Bufo marinus*, *B. paracnemis* and *Rana* spp. (see Bentley 1958, 1966a, 1967a,b; Aceves et al. 1970; Coviello 1969, 1970; Levine et al. 1976; Davis and Goodman 1986; Mia et al. 1987; Henderson and Kime 1987). Following treatment with vasopressin, intramembrane particles (IMP) appear as aggregates in the luminal membrane (Muller et al. 1980; Rapoprot et al. 1981). This seems to be a response of the bladder epithelium to changes in permeability to water (Brown et al. 1983). It has also been shown that microtubules and microfilaments play a role in the apical membrane permeability to water elicited by vasopressin (Pearl and Taylor 1985).

In the aquatic ranids, the rate of electrolyte and water excretion from the bladder is higher when the animal is dehydrated (Sinsch 1991).

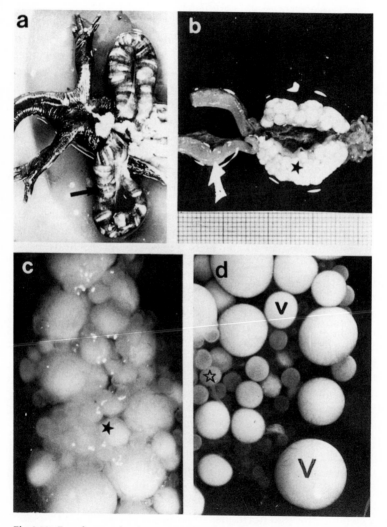

Fig. 3.13. Female reproductive system of *Salamandra* in situ showing the uterus filled with larvae (*arrow*; **a**, ×2), and the excised reproductive tract showing the ovaries (*asterisk*) and the uterus (*arrow*; **b**, ×2). The ovary is wrapped up in a transparent membrane through which the oocytes are visible (**c**, ×6). After cutting through the membrane, large and small vitellogenic oocytes (*V*, *v*) and previtellogenic oocytes (*open asterisk*) can be seen, counted and measured (**d**, ×6)

Ultrastructural localization of K^+-dependent p-nitrophenyl-phosphatase activity in the urinary bladder of *Salamandra salamandra* has shown the presence of reaction products on the basolateral membrane of the granular cells and the MRC (de Piceis Polver et al. 1985). Vasopressin induces the formation of microvilli on the apical surfaces of the granular cells (Le Furgey and Tisher 1981).

3.5
Structure and Function of the Female Reproductive System

3.5.1
The Ovary and Oogenesis

Several studies dealt with reproduction in *Salamandra*, some of which were reviewed recently (Greven and Guex 1994; Joly et al. 1994). The first detailed study of the course of reproduction of a xeric-inhabiting urodele involved *Salamandra salamandra infraimmaculata* (Sharon 1995). This urodele has a palaearctic distribution, but the population of that particular subspecies is entirely confined to mountainous habitats within northern Israel. Their habitats are located in three areas within the Mediterranean region. Since these areas are isolated, they are inhabited by disjunct populations of salamanders (Warburg 1986a, 1994). In these fringe habitats salamanders are extremely limited by the availability of water for breeding purposes even during their breeding season in the winter. On the other hand, in Europe water is available for breeding during their entire breeding season in the spring and summer. However, oogenesis has not been described in any detail so far, and no quantitative analysis has been undertaken throughout the salamander's activity period.

In her study Sharon (1995) found the ovaries to be wrapped in a thin, transparent membrane which protected the oocytes. These were attached to the inner side of the ovaries (Fig. 3.13). The average number of ooctyes was 218 ± 56.5 (ranging between 108 and 407).

The ovarian mass was significantly correlated with body mass (Jorgensen et al. 1979). This relationship was described by an allometric equation in *Bufo viridis* in which ovarian mass $= 0.084 \times$ (body mass) $1.23\,g$ (Jorgensen 1984). There was also a significant correlation between ovarian length and both total and pre-vitellogenic oocyte numbers. In *Salamandra*, the ovarian mass and length did not change significantly during the reproductive cycle, unlike in *Bufo cognatus* where the ovaries undergo annual variation (Long 1987).

Three stages in oogenesis are recognized: pre-vitellogenic, vitellogenic and post-vitellogenic. The first two stages were present in all ovaries. Average oocyte diameter was $3.19 \pm 0.29\,mm$ (ranging between 2.0 and 3.9 mm). The ovaries of all females contained pre-vitellogenic and vitellogenic oocytes. The ovary did not change in mass or length throughout the year. There was no correlation between ovarian mass and number of oocytes. The average number of pre-vitellogenic oocytes in the ovary was 140.47 ± 50.49, and the average number of vitellogenic oocytes was 75.59 ± 19.74 (34.86% of the total oocyte number).

The percentage of oocytes reaching the vitellogenic stage was constant, independent of season and reproductive stage.

3.5.1.1
Oogenesis

Oogenesis is the process whereby oogonia that multiply by mitosis are transformed into oocytes. In Amphibia in general during the pre-vitellogenic phase, Lofts recognized a primary and a secondary phase characterized by the accumulation of yolk and growth. Yolk deposition took place during aestivation and migration to the breeding sites (in *Taricha*) 5–6 months before ovulation (Miller and Robbins 1954). In *Salamandra* the dimensions of yolk platelets increased with egg development (Grodzinski 1975), and consequently the egg diameter was 5 mm (Joly and Picheral 1972). The post-ovulatory oocytes of *Salamandra* were larger and persist for a longer duration (Saidapur 1982).

Ovulation in *Salamanadra* has been described only once (Joly 1986). The process of ovulation is extremely rapid, and thus difficult to observe and record. Moreover, since this species is rather rare and protected, it was not possible to gather larger samples at any one time, especialy as it disappears for about half of the year into hiding places where it cannot be reached.

3.5.2
Mating, Sperm Storage, Fertilization and Gestation

Apparently, the female salamander can uncouple mating, ovulation and fertilization (Greven and Guex 1994). Joly et al. (1994) suggested that the European salamanders may mate during any month of the year. All females were found to contain spermatozoa. This is unlikely to occur in *Salamandra salamandra infraimmaculata* as the genders meet near the ponds only during a short period in the winter.

During amplexus a spermatophore is absorbed into the female's cloaca. It finds its way into a special receptaculum seminis (spermatheca) consisting of a complex of tubules (Siebold 1858). This tubuloalveolar exocrine gland reacts positively to both AB and PAS (Sever 1994).

Sever (1991, 1992, 1993) and Joly et al. (1994) found that spermatheca always contained spermatozoa, and that sperm was preserved there for more than 2 years (Boisseau and Joly 1975). On the other hand, Sever et al. (1995) found that the sperm in *Ambystoma* does not persist longer than 6 months. It appears to undergo degradation (Brizzi et al. 1995). In that salamander, multiple mating and possibly sperm competition took place.

Fertilization in *Ambystoma* took place during oviposition (Sever et al. 1995). Schwalbe (1896) never actually found spermatozoa in the oviduct or in the spermatheca of female salamander with an empty uterus and mature oocytes. Since in *Salamandra* spermatozoa cannot pass a uterus full with larvae, he concluded that fertilization must inevitably take place in the oviduct, as the eggs found in the oviduct were already fertilized by the end of May. According to Joly et al. (1994), the eggs are fertilized during ovulation while passing through the glandular part of the oviduct.

In *S. s. infraimmaculata*, ovulation and fertilization must take place sometime after the larvae have been released in the late autumn and beginning of the winter (November-January) and prior to the spring (March–April) when mature oocytes were still present in the ovaries. In view of this, gestation in *S. s. infraimmaculata* lasts between 4 and 5 months.

Gestation in *Salamandra* in Europe lasted 1 year (Wunderer 1910) or 4–12 months (Joly and Picheral 1972). A few salamanders breed twice yearly, some of them without mating a second time (Schwalbe 1896; Joly et al. 1994). In a long-term study on salamanders on Mt Carmel, we found that individual females bred during subsequent years. Our records have shown that some females bred even 5 years in a row (Warburg 1992b, and in prep.). This strengthens our conclusion that gestation cannot last longer than 5-6 months. In some of the female newt (*Taricha*) population no mature ovaries were observed, thereby indicating that some females of that species did not breed every year (Miller and Robbins 1954).

3.5.3
Conclusions

In this chapter I have tried to describe the present state of knowledge in some areas concerning the structure and function of some organs and systems. These were selected because of their particular importance in understanding the ecophysiological adaptations of amphibians inhabiting xeric environments. Our state of knowledge in these areas is far from satistactory. The main reason is the fact that most descriptive research was done on the more common European or American amphibians, especially those that could be bred in the laboratory. Consequently, our knowledge of the many xeric-adapted amphibians is still rather fragmentary. The main approach in future research in that field should be to try and combine a study of structure and function. What are the changes in transport phenomena as related to structural changes of the skin (due to ontogenesis, or phase shifting from aquatic to terrestrial and vice versa) or changes in the role played by the skin, gills or lungs in gas exchange (respiration or CO_2 elimination) during ontogenesis or under experimentally controlled ambient conditions. Likewise, our knowledge concerning changes in kidney structure from pronephros to mesonephros during ontogenesis or under different water shortage or salt loading conditions is still rather limited.

CHAPTER 4

Receptors, Perception and Behavioural Responses

4.1
Photoreception and the Response to Light

The visual cells of anurans and urodeles are similar. There are two types of rods, green and red, and a single cone with yellow sensitivity (Himstedt 1982; King et al. 1994).

Amphibians can distinguish wavelengths between 760 mμ and 400 mμ (Kasperczyk 1971). However, they vary in their type of colour vision. Thus, some Pelobatidae and Bufonidae are completely blind to colours. *Hyla arborea* seems to distinguish between blue and blue-green colours, whereas *Salamandra salamandra* is capable of distinguishing between any colour and between different shades of grey (Birukow 1950). This urodele was also found to be sensitive to red and violet as well as to yellow, yellow-green (Kasperczyk 1971; Himstedt 1973), blue-green and green-red (Przyrembel et al. 1995), and blue (Tempel et al. 1982). In *Pelobates*, vision changes during metamorphosis, and the spadefoot toads become more sensitive to yellow than to red (Hödl 1975). Adult *Salamandra* have a lower threshold to light than larvae do (Himstedt 1973).

Larval *Salamandra* have a dioptric apparatus enabling them to see under water (Manteuffel et al. 1977). The larval lens is more spherical than, and in close proximity to, the flat cornea (Sivak and Warburg 1980). Subsequently, during metamorphosis, the lens flattens, and the cornea becomes more concave. In *Pelobates syriacus*, in addition to these changes, the eye assumes a more rostral position, and the post-metamorphic stages have more binocular vision (Sivak and Warburg 1983).

Eyeless toads (*Bufo* spp.) and salamanders (*Ambystoma* spp.) retain the ability to orient (Heusser 1960; Taylor 1972 respectively). Both the pineal organ and the pineal body are sensitive to light wavelength and intensity and can be considered as effective extra-ocular photoreceptors (Adler 1970; Taylor and Adler 1978). The role of the pineal gland in regulating behavioural responses will be discussed later in this chapter. In addition, wavelength-sensitive photoreceptors are known from the skin of some Amphibia (*Ambystoma* larvae, Becker and Cone 1966).

The selection of darkness (= skototaxis) often occurs in amphibians. *Salamandra* is strongly negatively phototactic (Tempel et al. 1982). Juvenile salamanders showed an increase in contrast sensitivity. They are more sensitive to dark objects than to bright ones (Manteuffel and Himstedt 1986). On the other hand, *Ambystoma* larvae are photopositive (Schneider 1968; Anderson 1972). *Bufo fowleri* was found to be strongly positively phototactic (Martof 1962a). This response was much less pronounced at high temperatures (37 °C), and the response time was at least doubled.

Moisture content was shown to affect normal phototactic behaviour in urodeles (Griffiths 1984).

Jaeger and Hailman (1973) found that 87% of the many anuran species they have studied (among them pelobatids: *Scaphiopus* spp., bufonids: *Bufo punctatus*, and hylids: *Litoria caerulea*) have shown photopositive responses in the laboratory; only 8% were photonegative. Most species show a response to the blue part of the spectrum (Hailman and Jaeger 1974). Schneider (1968) described the tendency of larval *Ambystoma* to swim to an illuminated area. On the other hand, most of the species studied showed evidence of nocturnal behaviour (Hailman 1984).

A change appears to take place in the phototactic response of amphibians during ontogenesis. Thus, adult salamanders (*Ambystoma* spp.) show negative phototactic behaviour, whereas in some species larvae are positively phototactic (Anderson 1972; de Neff and Sever 1977). Larvae of urodeles show vertical migration correlated with light intensity (Anderson and Graham 1967).

4.2
Acoustic Behaviour

Four types of calls can be distinguished in *Hyla arborea savignyi*: mating, territorial, solitary and distress (Schneider and Nevo 1972). Call duration, intervals between calls and the number of pulses decrease with rising temperature. The choruses of male frogs began at light intensities below 260 Lux and air temperature above 8 °C (Schneider 1971). The main function of the male's call is to attract the females during the breeding season by advertising (Brzoska and Schneider 1982). The burrowing toad, *Heleioporus*, calls from its burrow. The call is transmitted by resonance (Bailey and Roberts 1981). Salamanders lacking a middle ear cavity and a tympanum can nevertheless sense vibrations (Himstead 1994). One such mechanism could be due to the displacement of fluid in the lymphatic system of *Salamandra* (Smith 1968). In the aquatic larvae of *Ambystoma tigrinum*, the air-filled mouth cavity plays a role in transmitting pulsations produced by pressure fluctuations as caused by underwater sound (Hetherington and Lombard 1983).

4.3
Olfaction

Ambystoma tigrinum is capable of discriminating between odours (Mason et al. 1980, 1981; Mason and Stevens 1981). Both larvae and adults are capable of detecting food through olfaction (Nicholas 1922). When the olfactory tract was severed, *Bufo fowleri* was unable to detect odours (Martof 1962b). Finally, female newts when treated with prolactin and gonadotropins released a male-attracting, water-soluble substance of 5000 daltons molecular weight (Matsuda et al. 1994).

4.4
Selection of Suitable Substrate Moisture and pH

Many anurans burrow into the ground. These include bufonids, pelobatids, myobatrachids and leptodacylids. In *Bufo americanus* the response to substrate moisture is noticeable under stressful thermal conditions (Bundy and Tracy 1977).

The nature of the substrate, its texture and moisture content release the burrowing response in *Pelobates* (Meissner 1970). Optical, olfactory and tactile stimuli are also of importance in locating a suitable site for burrowing. In the *Scaphiopus* post-metamorphic stage, the selection of a retreat is largely influenced by moisture conditions. Even moist cattle dung serves as a suitable hiding place under which young toadlets will burrow (Weintraub 1980). In *Bufo viridis*, increased temperature causes an increase in the time spent beneath the ground surface (Hoffman and Katz 1989).

Digging by *Scaphiopus* is intermittent as it requires a large amount of energy and has been found to increase the rate of oxygen consumption

Fig. 4.1. Moisture responses in juvenile *Salamandra*. A choice of sand and soil at two moistures: 87.5% (*top*) and 62.5% (*bottom*) was offered. The animals "preferred" the high soil moisture

(Seymour 1973a,b,c). Salamanders usually utilize other animals' burrows or hide in crevices or under stones. *Ambystoma tigrinum* is, however, capable of digging its own burrows (Semlitsch 1983a).

Among dehydrated *Ambystoma*, substrate moisture is a significant cue in selecting a shelter (Marangio and Anderson 1977). Similarly, in the terrestrial phase of the newt *Triturus vittatus*, selection of a suitable hiding place is largely due to the soil's moisture content (Degani 1982b, 1984b). Juvenile salamanders (*S. salamandra*) respond significantly to higher soil moisture content (Fig. 4.1; Degani and Warburg 1980; Degani and Mendelssohn 1981b). Apparently, the adults are much less dependent on soil moisture in their response. Thyroxine treatment enhances the selection of substrate moisture in the post-metamorphic stage of *Ambystoma tigrinum* (Duvall and Norris 1980).

Low-frequency sounds are the main cue for the emergence of *Scaphiopus* from its underground aestivation, but a slight rainfall elicited emergence from shallow burrows (Dimmit and Ruibal 1980b).

Burrowing is triggered by a flight reaction from a predator in the aquatic neotenic *Ambystoma* (Taylor 1983).

Ambystoma spp. are capable of selecting a substrate of suitable pH (Mushinsky and Brodie 1975).

4.5
Feeding Behaviour

Salamanders respond to both moving and stationary objects (Himstedt et al. 1978). The prey must move at a certain velocity in order to elicit prey-catching behaviour (Luthardt and Roth 1979a). Adult *Ambystoma* spp. assume a sit-and-wait feeding strategy. Visual stimuli act as their main cues for prey capture (Lindquist and Bachmann 1980, 1982). Salamander larvae have two main feeding methods: snapping at organisms and gulping water, followed by sieving out the smaller organisms (Dineen 1955).

On the other hand, chemical cues are significant for prey recognition. In the salamander, *S. salamandra*, prey catching responses are stimulated by odour rather than by visual cues (Luthardt and Roth 1983). There seems to be a change in the response during metamorphosis (Himstedt et al. 1976). Thus, the shape of the body (prey) stimulates larvae in a different manner than it does adult salamanders (Himstedt et al. 1976). As a result, elongation of the prey facilitates prey catching by *Bufo marinus* (Ingle and McKinley 1978). *Batrachoseps* prefers larger-sized prey and tends to ignore the smaller ones (Maiorana 1978). Fourteen different components of feeding behaviour have been recognized by Lindquist and Bachmann (1980).

4.6
Locomotory Rhythms and Activity Patterns

In *Bufo regularis* activity has been found to increase with temperature (Cloudsley-Thompson 1967): an increase in temperature up to 38 °C doubled or tripled the activity (Higginbotham 1939). In the aquatic stages of *Ambystoma tigrinum*, temperatures above 10 °C triggered activity (Whitford and Massey 1970). In salamanders (*Salamandra*) activity was highest at low temperatures (10 °C) and high relative humidity (90%); Degani and Mendelssohn 1981b). Similarly, in bufonids, humidity modified the activity pattern. Thus, in *Bufo americanus*, both body temperature and hydration state affected locomotory activity (Preest and Pough 1989) and in *Bufo boreas*, dehydration led to increased activity (Putnam and Hillman 1977).

Diel rhythmicity is more pronounced in the larvae than in the early post-metamorphic stage of *Triturus* (Himstedt 1971). However, adult *Salamandra* and *Triturus* spp. show a marked rhythmic activity. In *Salamandra* the peaks of activity are at dusk and dawn. In *Bufo boreas* there is a bimodal pattern of nocturnal activity peaking between 20:30–24:00 and 02:00–04:00 (Hailman 1984).

The pineal organ may play a role in the entrainment of the diurnal rhythm in amphibians (Adler 1971). The endogenous circadian rhythm of melatonin levels may be directly involved in synchronizing the circadian locomotor rhythm in *Cynops pyrrhogaster* (Chiba et al. 1995).

4.7
Orientation, Homing and Migration

4.7.1
Orientation

Celestial compass orientation developed early in anuran history (Ferguson and Landreth 1966; Ferguson 1967). It is known to be of importance in some bufonids and hylids, as well as in urodeles.

Bufo marinus can home using visual cues or topographic landmarks to aid in finding the way (Brattstrom 1962, 1963). A local topography is memorized by young post-metamorphs of *Bufo japonicus* during their first trip (Ishii et al. 1995). The learning and memory processes are related to MSH levels in the brain (Kim et al. 1995). Orientation was more direct under clear skies (for *Bufo boreas*, see Tracy and Dole 1969b). Breeding adults of *Bufo fowleri* use sun compass orientation, whereas the non-breeding adults use topographic cues (Ferguson 1967). Where topography is interrupted and visual cues are therefore of limited value, *Bufo fowleri* uses geotaxis for orientation (Fitzgerald and Bider 1974).*Salamandra* is capable of using visual cues as well as celestial compass orientation for its homing (Himstedt and Plasa 1979; Plasa 1979). There is evidence of use of the polarization pattern of the sky for orientation in newts and frogs (Taylor 1972). On the other hand, there is no evidence for celestial compass orientation in adult *Bufo valliceps* (Grubb 1973a), but juvenile toadlets used a combination of sun compass and topographic cues (Grubb 1973b). Apparently, directional movements are affected by celestial cues, whereas topographic landmarks provide the main information enabling the evaluation of the distance to the breeding site during the migration of *Ambystoma* (Douglas 1979, 1981). In *Salamandra*, orientation during homing was guided by visual landmarks at very low light intensities (Himstead 1994). On the other hand, *Ambystoma tigrinum* and *A. texanum* larvae show a non-random orientation (Blaustein and Walls 1995). Extra-optic photoreception probably plays a role in orientation, as in *Bufo* eyes are not essential (Heusser 1960). Olfaction does not seem to be essential to the orientation of *Bufo americanus* (Oldham 1966).

4.7.2
Homing

The tendency of individuals to return to their home areas following displacement is called site tenacity. We have unambiguous evidence that both male and breeding female *Salamandra* return to their breeding ponds year after year (Warburg 1996; Fig. 4.1). They are largely guided by visual landmarks (Himstedt 1994).

Bufo punctatus homes successfully if displaced to a distance of 900 m (Weintraub 1974), whereas *B. boreas* can only do so for a distance of 200 m (Tracy and Dole 1969a). The speed of travel of *B. boreas* is temperature-dependent and increases with rising body temperature (Putnam and Bennett 1981). *Bufo viridis* is capable of finding gaps in a drift fence (Collett 1982). *Ambystoma* spp. have been observed entering and leaving by the same routes (Stenhouse 1985).

Visual cues are used for homing orientation in *B. valliceps*, together with olfactory cues (Grubb 1970, 1973b, 1976). Because blinded *B. boreas* toads were able to orient, Tracy and Dole (1969a,b) concluded that olfactory cues must be the principal means of orientation.

4.7.3
Migration and Breeding Behaviour

The movement to breeding sites (or "water drive") was described many years ago mainly in urodeles (Twitty 1959; Grant 1961). In *Ambystoma* spp., this migration takes place mainly at night, possibly to avoid predation (Semlitsch and Pechmann 1985). Migrating species differ in their diel patterns of migration.

Migration to the breeding site is triggered by a combination of rainfall and temperature. In *Ambystoma talpoideum* and in *Salamandra*, breeding migration is initiated by both rainfall and temperature (Hardy and Raymond 1980;

Warburg 1986a,b). The pattern and amount of rainfall are both of significance (Gibbons and Bennett 1974). Apparently, temperature is of greater significance than rainfall in initiating breeding in *Ambystoma* (Semlitsch 1985a). Movement towards breeding sites occurs along well-defined routes (Semlitsch 1981). Breeding migration in *Ambystoma maculatum* is correlated with soil temperature (Sexton et al. 1990).

The timing of appearance at the breeding sites as well as the duration of time spent there are of major significance for successful breeding. Generally, in salamanders the males appear first and stay longer at the breeding sites (Douglas 1979; Hardy and Raymond 1980; Warburg 1986a,b; Semlitsch et al. 1993). There is a difference even among closely related *Ambystoma* species in their diel pattern of migratory activity (Semlitsch and Pechmann 1985). Prolactin appears to have an effect on sexual behaviour during breeding in *Triturus* (Giorgio et al. 1982). This effect is probably due to the stimulation of the salamanders to seek and remain in water (Moriya 1982; Moriya and Dent 1986).

A mass migration of thousands of toads (*Bufo boreas*) to their breeding site was observed by Tracy and Dole (1969a,b). Several examples given in Rose (1962) are discussed in Chapter 6. In *Ambystoma* spp. the animals appear to come from the same retreat year after year (Stenhouse 1985).

Male anurans engaged in explosive breeding tend to aggregate and scramble for mates. On the other hand, males with a prolonged breeding period call from stationary posts (Wells 1977).

4.8
Social Behaviour

4.8.1
Aggregation Behaviour

This subject has been recently reviewed by Blaustein and Walls (1995).

The aquatic larval stages of *Scaphiopus* spp. as well as the larvae of other anuran and urodele species are known to aggregate in groups (Bragg 1954, 1959, 1960; Bragg and Brook 1958; Bragg and King 1960; Wassersug 1973; Blaustein and Waldman 1992). It is largely the bottom-dwelling species that show this character. Thus, both *Bufo carens* and *Pyxicephalus adspesurs* have this behavioural trait (Van Dijk 1972). Aggregation is less apparent when ponds remain full for longer periods (Bragg 1959). In addition, larvae are known to form "schools" during feeding (Wassersug 1973). These "schools" are clusters of polarized individual tadpoles facing the same direction and remaining much of the time in physical contact with each other, as observed in *Bufo, Scaphiopus* and *Xenopus* tadpoles (Wassersug 1973). Thus, tadpoles of *Bufo woodhousi* forming such "schools" orient themselves parallel to the nearest neighbour (Wassersug et al. 1981).

Some individuals within these schools eventually die and are cannibalized by their siblings.

Newly metamorphosed *Bufo americanus* form dense aggregations in the periphery of the ponds in response to the risk of desiccation (Heinen 1993). Similar observations were made on young *Salamandra* metamorphs aggregating under stones near the pond.

The phenomenon of aggregation may have a selective advantage in providing food (through cannibalism) for some members of the cohort who may consequently survive to metamorphose. Cannibalism is more pronounced in denser populations of larvae (Degani et al. 1980; Pfennig and Collins 1993). In salamanders (*Salamandra*), cannibalism involves members of the same brood (Degani et al. 1982). On the other hand, there is evidence in *Scaphiopus* that this cannibalism is less pronounced among siblings than unrelated other members of that species (Pfennig et al. 1993).

4.8.2
Kin Recognition

When reared in sibling groups during their early development. *Bufo americanus* show kin recognition (Waldman and Adler 1979; Waldman 1981, see reviewed by Waldman 1991; Blaustein and Walls 1995). Sixty-five percent of a school were biased in favour of their siblings (Waldman 1982). Likewise, in *Ambstoma opacum*, larvae were capable of discriminating siblings from non-siblings (Walls and Roudebush 1991). *Bufo boreas* tadpoles reared in mixed groups of siblings with other non-related larvae did not always show kinship recognition, although early experience affected this behaviour (O'Hara and Blaustein 1982, Walls and Blaustein 1994). There are some reports that kin recognition ability was retained throughout metamorphosis (Walls 1991; Blaustein and Waldman 1992).

Kinship cues are chemicals that are perceived by olfaction; they fade with time (Waldman 1985). The ability to perceive the signal diminishes after a certain stage of development (Blaustein et al. 1993). Pfennig et al. (1993) have noted that in *Scaphiopuds bombifrons* cannibalistic tadpoles nipped at conspecific tadpoles but ate only non-siblings. However, in *Ambystoma tigrinum nebulosus* cannibalistic larvae consumed their conspecifics (Pfennig and Collins 1993).

4.9
Endocrine Control of Behaviour

4.9.1
The Pineal Gland

The pineal gland is a diverticulum of the dorsal brain containing rod-like photoreceptors (Korf and Oksche 1986). It plays a role in controlling photoperiodism and reproduction. Pineal hormones include melatonin which is synthesized from serotonin and suppresses gonad development during short days and under constant light (Binkley 1979). In *Ambystoma*, plasma melatonin levels increase significantly during the night (Gern and Norris 1979). Likewise, melanocyte stimulating hormone (MSH) peaked during 2100–0500 h in *Bufo speciosus* (Kim et al. 1995). On the other hand, high temperature causes a reduction in circulating melatonin in the plasma of *Ambystoma* (Gern et al. 1983). Treating urodeles (*Cynops pyrrhogaster*) experimentally with melatonin caused entrainment of their circadian rhythms (Kikuchi et al. 1989; Chiba et al. 1995).

4.9.2
Courtship and Mating Behaviour and Their Endocrine Control

The male *Ambystoma gracile* captures the female using his hind limbs (Licht 1969). In *Salamandra*, the male lies beneath the female whilst mating, holding her in amplexus with his forelimbs (Joly 1966). Olfactory, optical and tactile stimuli play a role in this behavioural sequence.

Arginine, vasotocin, prolactin and gonadotropin elicit mating behaviour in newts (Zoeller and Moore 1982, 1988; Toyoda and Kikuyama 1990a,b, 1995). In *Bufo*, amplexus induces a surge of gonadotropin during the breeding season (Itoh and Ishii 1990). On the other hand, injecting *Cynops pyrrhogaster* with antiserum against PRL caused a decline in courtship behaviour (Toyoda et al. 1994). AVT appears to cause an increase in call rates in male *Rana catesbeiana* but not in females; during spring but not during fall (Boyd 1992,1994a). AVT immunoreactive cells were localized in six different regions of the brain (Boyd 1994b), and their density was significantly greater in males (Boyd et al. 1992).

4.10
Learning

Bufo boreas can learn to distinguish between the odours of different prey (Dole at al. 1981). Furthermore, *B. terrestris* is capable of associating unpalatable prey with a certain background colour and previous experience by *Salamandra* affected its prey capture (Luthardt and Roth 1979b; Roth and Luthardt 1980).

Post-metamorphic salamanders (*Ambystoma*) are capable of associating an optical cue with an acoustical stimulus (Ray 1970). Thus when a vibration sounded during bright light, the salamanders learned to associate the sound with the light and avoid it.

4.11
Conclusions

In conclusion, amphibians form xeric habitats appear to possess similar behavioural patterns. There is no proof that their response to moisture is more pronounced, nor that they are more efficient in their other behavioural responses. Does their response to precipitation and temperature elicit breeding migration in a more pronounced way than it does in other amphibians? In this case, time may be more critical since arriving early at the ponds may be crucial for the eventual success of the larvae in completing metamorphosis. There is no experimental evidence on any of these subjects.

Since the larvae as aquatic animals have quite a different life to that of the terrestrial adult, does it mean that previous experience by the larva will have no effect on the subsequent behaviour of the adult? This interesting question, relevant to all animals with complex life histories, has so far not received its appropriate experimental study.

CHAPTER 5

Physiological Adaptations

5.1
Water Balance

Water balance in amphibians is maintained by matching water loss and up-take. Water is lost via the skin and lungs during respiration, and via the kidney and bladder. Loss of water through respiration may be negligible since inhibition of pulmonary respiration does not significantly alter water loss (Bentley and Yorio 1979).

There are a number of reviews on the subject of water balance which can be consulted for further details (Deyrup 1964; Scheer et al. 1974; Shoemaker and Nagy 1977; Alvarado 1979; Shoemaker 1987, 1988; Warburg 1972, 1988; Warburg and Rosenberg 1990; Shoemaker et al. 1992; see also Table 5.1).

5.1.1
Water Content

Water makes up around 80% of an amphibian's body weight (Deyrup 1964; Bentley 1966a). Thus, amphibians have the highest proportion of body water among vertebrates (Brown and Brown 1987). The water content of anurans is 79.2% (Claussen 1969), whereas that of urodeles is 78.1% (Alvarado 1979). A few data have been compiled in Table 5.2. In the various amphibian species whose water content has been studied, it has been found to range between 70 and 86%. Some of these calculations are presented as percentage of body weight, others are based on dry weight (Brown et al. 1986). Moreover, the physiological state of an animal affects its water content, and this factor is not usually taken into consideration. Thus, no generalization based on data obtained in these different ways is possible. There is, however, no significant difference in the water content of xeric-inhabiting amphibians compared with that of mesic or aquatic species (see Table 1 in Deyrup 1964).

5.1.2
Remarks about Techniques

In most studies evaporative water loss (EWL) and the vital limit of desiccation till death have been measured in dry, moving air. Under those conditions the survival limits were reached much sooner than when frogs were desiccated in

Table 5.1. Anuran species from xeric habitats in which some aspect of water economy was studied

Species	Reference
Bufonidae	
Bufo viridis	Gordon (1962); Warburg (1971b, 1972); Mack and Hanke (1977a,b); Khlebovich and Velikanov (1982); Goldenberg and Warburg (1983); Degani et al. (1984)
B. punctatus	McClanahan and Baldwin (1969); Claussen (1969); Fair (1970)
B. cognatus	Ruibal (1962); McClanahan (1964); Schmid (1965b); McClanahan and Baldwin (1969)
B. regularis	Ewer (1951, 1952a,b); Cloudsley-Thompson (1967)
B. woodhousei	Spight (1968); Jones (1982)
Hylidae	
Litoria rubella	Main and Bentley (1964); Warburg (1965, 1967)
L. caerulea	Main and Bentley (1964); Warburg (1967)
L. latopalmata	Main and Bentley (1964)
L. moorei	Main and Bentley (1964)
Hyla arborea	Overton (1904); Reichling (1958); Warburg (1971b, 1972); Degani and Warburg (1984)
H. cadaverina	Jones (1982)
Phyllomedusa sauvagei	Shoemaker et al. (1972); Shoemaker and McClanahan (1975); McClanahan et al. (1978)
Cyclorana mainii	Warburg (1967)
C. platycephalus	Main and Bentley (1964)
Hyperoliidae	
Hyperolius nasutus	Withers et al. (1982a,b)
H. marmoratus	Loveridge (1976); Withers et al. (1982a)
H. viridiflavus	Kobelt and Linsenmair (1986); Geise and Linsenmair (1986)
Pelobatidae	
Pelobates syriacus	Warburg (1971b); Degani et al. (1983); Goldenberg and Warburg (1983); Degani and Nevo (1986)
Scaphiopus multiplicatus	Jones (1980a,b)
S. hammondi	Thorson and Svihla (1943); Thorson (1955); Shoemaker et al. (1969); Ruibal et al. (1969); Lasiewski and Bartholomew (1969)
S. couchi	Thorson and Svihla (1943); McClanahan (1964, 1967, 1972); Mayhew (1965); Shoemaker et al. (1969); McClanahan and Baldwin (1969); Claussen (1969); Lasiewski and Bartholomew (1969); Hillyard (1976a,b); Jones (1978,1980a,b); Hillman (1980)
S. holbrooki	Thorson and Svihla (1943)
Leptodactylidae	
Eleutherodactylus coqui	van Berkum et al. (1982); Pough et al. (1983); Taigen et al. (1984)
E. portoricensis	Heatwole et al. (1969)
Lepidobatrachus llanensis	McClanahan et al. (1976, 1983)
Ceratophrys ornata	Canziani and Cannata (1980)
Myobatrachidae	
Notaden nichollsi	Main and Bentley (1964)
Neobatrachus pictus	Warburg (1965)
N. centralis	Bentley et al. (1958); Warburg (1965)
N. sutor	Bentley et al. (1958)
N. wilsmorei	Bentley et al. (1958)
N. pelobatoides	Bentley et al. (1958)
Heleioporus eyrei	Bentley et al. (1958); Bentley (1959); Packer (1963)
H. inornatus	Bentley et al. (1958)
H. psammophilus	Bentley et al. (1958)

Table 5.1 (*Contd.*)

Species	Reference
H. australiacus	Bentley et al. (1958); Lee (1968)
H. albopunctatus	Bentley et al. (1958)
H. barycragus	Main (1968)
Pseudophryne bibroni	Warburg (1965)
Limnodynastes ornatus	Warburg (1965)
L. dorsalis	Warburg (1965, 1967)
Rhacophoridae	
Chiromantis xerampelina	Loveridge (1970); Balinsky et al. (1976)
C. petersi	Drewes et al. (1977)
Urodela	
Salamandridae	
Triturus vittatus	Warburg (1971b, 1972); Goldenberg and Warburg (1983)
Salamandra salamandra	Warburg (1971b, 1972); Warburg and Degani (1979); Degani (1981a, 1982a); Goldenberg and Warburg (1983)
Ambystomatidae	
Ambystoma tigrinum	Alvarado (1972); Delson and Whitford (1973b)

Table 5.2. Water content (in percentage)

Family/species	Water content (%)	Reference
Anura		
Bufonidae		
Bufo marinus	78.0	Shoemaker (1964)
B. bufo	72.9–83.6	Zamachowsky (1977)
B. viridis	72.7	Hoffman and Katz (1991)
B. boreas	79.8	Thorson and Svihla (1943)
B. boreas	79.0	Cloudsley-Thompson (1967)
B. regularis	77.0	Cloudsley-Thompson (1967)
B. terrestris	78.8	Cloudsley-Thompson (1967)
Hylidae		
Litoria caerulea	79.4	Main and Bentley (1964)
L. latopalmata	75.2	Main and Bentley (1964)
L. rubella	77.6	Main and Bentley (1964)
L. moorei	71.2	Main and Bentley (1964)
Hyla versicolor	82.3	Farrell and MacMahon (1969)
Acris crepitans	77.2–78.9	Ralin and Rogers (1972)
Pseudacris streckeri	81.7–83.0	Ralin and Rogers (1972)
Cyclorana platycephalus	77.3	Main (1968)
Myobatrachidae		
Notaden nichollsi	83.0	Main (1968)
Heleioporus eyrei	81.7	Main (1968)
Pelobatidae		
Scaphiopus hammondi	80.0	Thorson and Svihla (1943)
S. holbrooki	79.5	Thorson and Svihla (1943)
Ranidae		
Rana arvalis	70.0–83.1	Zamachowski (1977)

still air since the rate of EWL is related to the speed of airflow (Warburg 1965; Claussen 1969). From an ecological point of view, still air is likely to surround most frogs (when they are out of water), as they usually seek hiding places in rock crevices, under stones or in burrows in the ground. In some hylids, hyperolids and rhacophorids that have been exposed to desiccation in open air, the lost water can normally be replenished without great effort. In some arboreal frogs, however, special anatomical and physiological adaptations enable an unusual degree of water conservation (as previously discussed in Chap. 3).

There appears to be general agreement among most researchers on the following points: (1) many terrestrial species of amphibians are capable of tolerating a loss of up to 50% of their body water; and (2) terrestrial adaptation among anurans is probably reflected less in their rate of water loss than in their rate of water uptake.

5.1.3
Evaporative Water Loss

EWL was studied by several authors (Table 5.3) and was found to be greater at higher temperatures than at lower ones (Warburg 1965; McClanahan et al.

Table 5.3. Evaporative water loss $(mg\,g^{-1}\,h^{-1})$

Species	EWL	Reference
Anura		
Bufonidae		
Bufo regularis	0.5–9[a]	Cloudsley-Thompson (1967, 1974)
B. cognatus	11.8–12.2	Shoemaker et al. (1972)
B. cognatus	51	Withers et al. (1984)
B. marinus	0.2–2.1[b]	Warburg (1965)
B. viridis	0.2–3.7[b]	Warburg (1971a)
B. mauritanicus	2.4–11.2[a]	Cloudsley-Thompson (1974)
Hylidae		
Hyla arenicolor	72	Withers et al. (1984)
H. arborea	12.9	Degani and Warburg (1984)
Litoria caerulea	55	Withers et al. (1984)
	0.25–2.81[b]	Warburg (1967)
L. rubella	0.8–3.2	Warburg (1965)
Cyclorana maini	0.45–3.04[b]	Warburg (1967)
Pachymedusa danicolor	10.4	Shoemaker and McClanahan (1975)
Agalychnis annae	11.6	Shoemaker and McClanahan (1975)
Phyllomedusa pailona	0.6	Shoemaker and McClanahan (1975)
P. azurae	1.5	Withers et al. (1984)
P. hypochondrialis	0.9	Shoemaker and McClanahan (1975)
P. sauvagei	0.6	Shoemaker and McClanahan (1975)
P. sauvagei	0.3–1.6	Shoemaker et al. (1972)
P. ihrenegi	0.6	Shoemaker et al. (1972)
Hyperolidae		
Hyperolius nasutus	4.5	Withers et al. (1982a)
H. marmorata	0.8	Withers et al. (1984)

Table 5.3 *(Contd.)*

Species	EWL	Reference
Myobatrachidae		
Neobatrachus pictus	0.8–1.7[b]	Warburg (1965)
N. centralis	1.7–2.7[b]	Warburg (1965)
Heleioporus eyrei	0.8	Packer (1963)
Limnodynastes dorsalis	0.5–4.4[b]	Warburg (1965)
L. ornatus	2.7	Warburg (1965)
Pelobatidae		
Pelobates syriacus	0.3–1.5[b]	Warburg (1971a)
Scaphiopus couchi	13.6–14.2	Shoemaker et al. (1972)
Rhacophoridae		
Chiromantis petersi	0.7	Withers et al. (1984)
C. rufescens	0.5	Withers et al. (1984)
C. xerampelina	0.4	Withers et al. (1984)
Urodela		
Salamandridae		
Salamandra salamandra	0.9–4.2[b]	Warburg (1971b)
Triturus vittatus	0.8–6.9[b]	Warburg (1971b)
Ambystomatidae		
Ambystoma tigrinum	1.8%/h	Alvarado (1972)
A. macrodactylum	8.5%/h	Alvarado (1967)

[a] Between 30–45 °C.
[b] Between 20–40 °C.

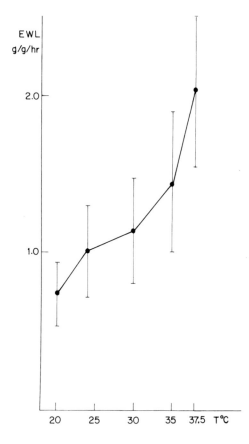

Fig. 5.1. Evaporative water loss (EWL) in *Litoria rubella* at different temperatures

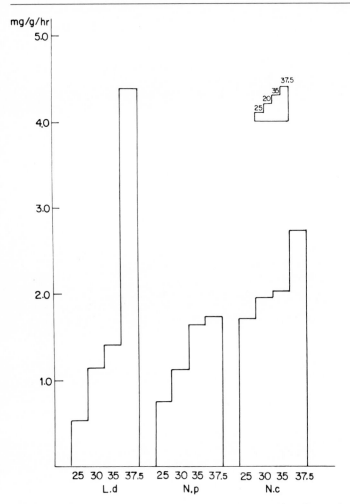

Fig. 5.2. EWL in three Australian frogs: *Limnodynastes dorsalis*, *Neobatrachus pictus* and *N. centralis* at four temperatures

1978; Warburg and Degani 1979; Figs. 5.1, 5.2). In the arboreal frogs *Litoria* spp. and *Phyllomedusa sauvagei*, EWL increased greatly at temperatures between 35 and 40 °C (McClanahan et al. 1978). Similarly, in the arboreal frogs *Chiromantis xerampelina* and *C. petersi*, EWL increased greatly when the temperature reached 38–39 °C (Drewes et al. 1977; Shoemaker et al. 1987). In the fossorial leptodactylid *Limnodynastes dorsalis*, this temperature effect was much less pronounced. In *Salamandra salamandra*, EWL increased greatly at 35 °C (Warburg and Degani 1979; Figs. 5.3, 5.4).

Moreover, water loss is greater in dry than in humid air. In *Bufo regularis* the EWL is related to the saturation deficit of the air (Cloudsley-Thompson 1967). At a vapour pressure deficit (VPD) of 9.2 mmHg, the EWL of several frog

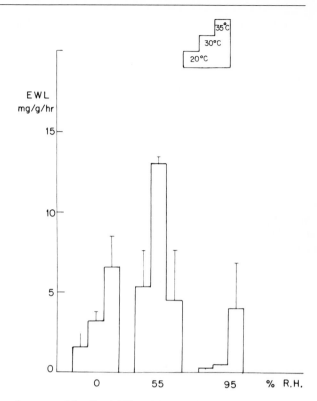

Fig. 5.3. EWL in adult *Salamandra* exposed for 4 h at different temperatures

species from mesic habitats followed the VPD curve up to 30 °C (Warburg 1965). The EWL was greater during the first 2–4 h of exposure in either dry or humid air (Figs. 5.5, 5.6). There is a steady rate of evaporation at lower temperatures. At higher temperatures, the EWL increases gradually. During long-term dehydration (in still air) for periods of either 8 or 3 days, most of the water is lost during the first period (2 h or 1st day). This is seen in some hylids: *Litoria caerulea*, *L. rubella*, and *Cyclorana maini* (Figs. 5.5, 5.6). *L. rubella* lost more water than other *Litoria* spp. at both 20 and 35 °C (Warburg 1967). This was true for temperatures up to 37.5 °C, but at 40 °C *Cyclorana maini* lost less water than the other hylids. Similar measurements were taken of *Ambystoma tigrinum* (Delson and Whitford 1973a,b). *Bufo viridis* lost 35.8% water during 10 h desiccation. This was replenished later when the toad was hydrated (Mack and Hanke 1977a). The rate of water uptake was higher in *Bufo viridis* than in *Rana* (Katz and Graham 1980).

There are specific differences in EWL among the various amphibian species (Table 5.3). Thorson (1955) found no relationship between EWL and adaptation to life on land. (He studied amphibians inhabiting mesic habitats in the State of Washington, some of which extend their distribution into semi-arid or

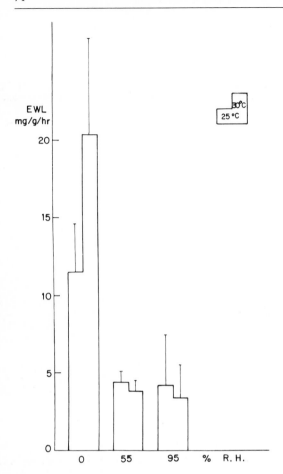

Fig. 5.4. EWL of juvenile *Sala-mandra* exposed for 4h at two temperatures

arid regions). In another study, no relationship between moisture conditions in the habitat and EWL could be established (Jameson 1966). Farrell and MacMahon (1969) arrived at a similar conclusion in their studies on some hylids.

The rate of water loss in the fossorial pelobatid *Pelobates syriacus* was comparatively lower than that of the hylid *Hyla arborea* or the bufonid *Bufo viridis* (Warburg 1971a). It is possible that the cranial ossification in *Pelobates* leads to a reduction in EWL (Seibert et al. 1974). The North American hylid *Hyla arenicolor* had comparatively low rates of EWL (Preest et al. 1992). Likewise, in the Australian hylids *Litoria fallax* and *L. peroni* (Amey and Grigg 1995) and some phyllomedusine (*Phyllomedusa* spp.) and hyperoliid (*Hyperolius* spp.) frogs the EWL was very low (Blaylock et al. 1976; Withers et al. 1982). The terrestrial phase of the urodeles *Triturus vittatus* and *S. salamandra* shows a higher rate of water loss than that of some anurans (Warburg 1971b).

Fig. 5.5. EWL of *Litoria rubella* at four temperatures

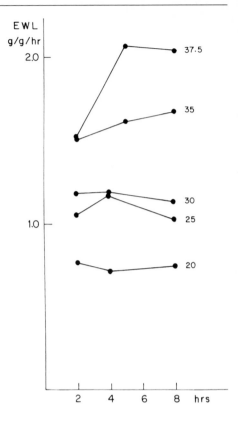

Juvenile amphibians usually lose more water and at a higher rate than adults. This is true of leptodactylids (*Limnodynastes dorsalis, Neobatrachus pictus*; Warburg 1965), pelobatids (*Pelobates syriacus*; Warburg 1971a), hylids (*Hyla arborea*; Degani and Warburg 1984), as well as urodeles such as *Ambystoma tigrinum* (Gehlbach et al. 1969) and *S. salamandra* (Warburg and Degani 1979; Figs. 5.3, 5.4).

There is a tendency among juvenile urodeles to aggregate in groups, especially in response to desiccation (Gehlbach et al. 1969; Warburg 1971b). Both anurans and urodeles are capable of conserving water better in groups than when exposed individually to dry air (Warburg 1971b; Warburg and Degani 1979). Thus, in *Ambystoma tigrinum*, body water loss was 1.8%/h, whereas in groups of 5, EWL was 0.9%/h (Alvarado 1967). Similarly, juvenile *Limnodynastes dorsalis* lowered their water loss by half when in groups of 5 (Johnson 1969b).

There are apparently seasonal differences in EWL. Thus, during the dry season the reed frog (*Hyperolius viridiflavus*) lost 1.2% body eight/day, whereas during the wet season EWL under similar experimental conditions, it was 30 times higher (Geise and Linsenmair 1986). Likewise, in *Litoria caerulea* the water flux was higher during the wet season (Christian and Green 1994).

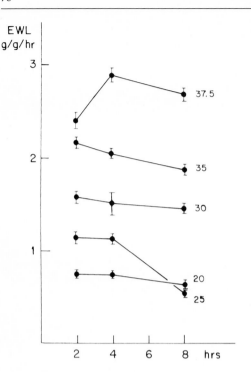

EWL
g/g/hr

3

2

1

37.5

35

30

20
25

2   4   6   8   hrs

Fig. 5.6. EWL in *Cyclorana maini* at four temperatures

Variations in the rate of water loss have been found between different isolated populations of *Salamandra salamandra* (Degani 1981b). *Litoria rubella* populations from different latitudes (in Australia) were found to respond differently in their rates of water loss in dry air (Warburg 1967). Thus, frogs of this species originating from Central Australia were more efficient water conservers than specimens belonging to the same species from either Western Australia or the interior of New South Wales. This may be a result of adaptation to the respective temperature ranges in each region. It was found experimentally that, to a certain extent, the acclimation temperatures of frogs affected their EWL (Figs. 5.5, 5.6). Thus, specimens of *Litoria rubella* acclimated at 30 °C lost less water than those acclimated at 10 °C (Warburg 1967). Differences in EWL were also found in two populations of a South American leptodactylid *Ceratophrys ornata*: from the arid Chaco and from the temperate region of Argentina. Members of the arid population were better water conservers (Canziani and Cannata 1980).

Activity levels greatly affected the rate of EWL in both *Bufo marinus* and *Litoria caerulea* (Heatwole and Newby 1972). When *Cyclorana maini* was allowed to burrow into moist sand at 20 °C, it lost weight (largely water) during a 10-day experimental period at a rate of $0.023\,\mathrm{mg\,g^{-1}\,h^{-1}}$ (Warburg 1967). In non-cocooned *Cyclorana platycephalus*, EWL was about $0.00260\,\mathrm{ml\,cm^{-2}\,h^{-1}}$ (van Beurden 1984).

5.1.4
Survival of Dehydration

Anurans did not die at 15% water loss (Schmid 1965b). Death occurred only when frogs were desiccated to about 50% of their body weight, depending on the species (Bentley et al. 1958; Bentley 1966a; Table 5.4). Thus, several species of *Heleioporus* and *Neobatrachus* survived up to 45% water loss. The fossorial leptodactylid *Heleioporus eyrei* was capable of losing about 22–45% of their body weight when in dry soil (Lee 1968). Several hylids survived between 32 and 45% loss of body weight (Main and Bentley 1964). The pelobatid *Scaphiopus couchi* survived a loss of 44.9% of its body water (Hillman 1980), whereas *S. holbrooki* survived 60% water loss (Schmid 1965b). The size of *S.*

Table 5.4. Critical water loss

Species	Water loss(%)	Reference
Anura		
Bufonidae		
Bufo americanus	54	Schmid (1969)
Hylidae		
Litoria moorei	31.4	Main and Bentley (1964)
L. caerulea	45.0	Main and Bentley (1964)
L. latopalmata	39.4	Main and Bentley (1964)
L. rubella	37.7	Main and Bentley (1964)
Acris crepitans	23.1–29.9	Ralin and Rogers (1972)
Pseudacris streckeri	44.3–44.8	Ralin and Rogers (1972)
Cyclorana platycephalus	38.5	Main and Bentley (1964)
Hyperolidae		
Hyperolius viridiflavus nitidulus	50	Geise and Linsenmair (1986)
Myobatrachidae		
Notaden nichollsi	38.5	Main and Bentley (1964)
N. bennetti	32.7	Heatwole et al. (1971)
Heleioporus psammophilus	41.3	Bentley et al. (1958)
H. inornatus	40.1	Bentley et al. (1958)
H. eyrei	40.7	Bentley et al. (1958)
H. australiensis	38.9	Bentley et al. (1958)
Neobatrachus sutor	44.1	Bentley et al. (1958)
N. pelobatoides	38.8	Bentley et al. (1958)
N. centralis	41.7	Bentley et al. (1958)
N. wilsmorei	43.9	Bentley et al. (1958)
Pelobatidae		
Scaphiopus couchi	50	McClanahan (1967)
Urodela		
Salamandridae		
Salamandra salamandra	40	Degani (1981a)
Ambystomatidae		
Ambystoma tigrinum	45	Alvarado (1972)

Table 5.5. Water uptake

Species	H$_2$O Uptake (mg mg^{-2} h^{-1})	Reference
Anura		
Bufonidae		
Bufo boreas	45.1 wt/h^{-1}	Shoemaker and McClanahan (1975)
Bufo boreas	41.4 wt/h^{-1}	Hillman (1980)
B. punctatus	62.3	McClanahan and Baldwin (1969)
B. viridis	75.8	Warburg (1971a)
B. cognatus	44.9	Hillman (1980)
B. regularis	50.0	Cloudsley-Thompson (1967)
Hylidae		
Hyla arborea	77.8	Warburg (1971a)
Litoria caerulea	16.0	Warburg (1965, 1967)
Litoria Caerulea	109.0	Main (1968)
L. rubella	11.0	Warburg (1965, 1967)
L. rubella	66.0	Main (1968)
Phyllomedusa sauvagei	51.2 wt/h^{-1}	Shoemaker and Bickler (1979)
P. sauvagei	35.0 wt/h^{-1}	Shoemaker and McClanahan (1975)
P. ihrenegi	24.0 wt/h^{-1}	Shoemaker and McClanahan (1975)
P. pailona	26.2 wt/h^{-1}	Shoemaker and McClanahan (1975)
P. hypochondrialis	29.1 wt/h^{-1}	Shoemaker and McClanahan (1975)
Agalychnis annae	38.0 wt/h^{-1}	Shoemaker and McClanahan (1975)
Cyclorana maini	17.0	Warburg (1965, 1967)
Cycloprana platycephalus	80	van Beurden (1984)
	92.0	Main (1968)
Hyperolidae		
Hyperolius viridiflavus taeniatus	20.4–37.7%/h	Geise and Linsenmair (1986, 1988)
Microhylidae		
Heleioporus psamophilus	44.0	Bentley et al. (1958)
H. inornatus	55.0	Bentley et al. (1958)
H. eyrei	52.5	Bentley et al. (1958)
H. australiacus	60.0	Bentley et al. (1958)
H. albopunctatus	57.7	Bentley et al. (1958)
Neobatrachus pelobatoides	33.3	Bentley et al. (1958)
N. centralis	55.7	Bentley et al. (1958)
N. sutor	84.8	Bentley et al. (1958)
N. wilsmorei	99.4	Bentley et al. (1958)
Notaden nichollsi	92.0	Main (1968)
Pelobatidae		
Pelobates syriacus	32.4	Warburg (1971a)
Rhcophoridae		
Rhacophorus spp.	38–42	Shoemaker and McClanahan (1980)
Chiromantis petersi	74.8% body wt/h	Drewes et al. (1977)
Urodela		
Salamandridae		
Salamandra salamandra	80.9 mg g^{-1} h^{-1}	Warburg (1971b)
Triturus vittatus (terr.)	104.2 mg g^{-1} h^{-1}	Warburg (1971b)

couchi metamorphs did not affect their dehydration tolerance (Newman and Dunham 1994). Among bufonids: *Bufo boreas* tolerated up to 41.4% body weight loss, *B. cognatus* 42.9% (Hillman 1980), *B. regularis* up to 50% (Cloudsley-Thompson 1967), *B. marinus* 52.6% (Krakauer 1970), and *B. americanus* up to 54% of their total body water (Schmid 1965b). Similarly, *Hyperolius viridiflavus* is capable of tolerating water loss up to 50% of its body weight (Geise and Linsenmair 1986). Among the urodeles, *Ambystoma opacum* survives dehydration up to 30% body water loss (Spight 1967a,b). Adult *Ambystoma tigrinum* survived between 35–55% body water loss (Alvarado 1972; Romspert and McClanhan 1981).

In fossorial anurans, water can be reabsorbed from the bladder as well as from the peritoneal cavity (Tables 5.5, 5.6). Among the bufonids, *B. arenarum* toads could store water in their bladder when dehydrated (Schmajuk and Segura 1982). Burrowing leptodactylid and myobatrachid frogs were capable of surviving long periods of dehydration by using dilute urine stored in their bladder (Warburg 1965). How is this done? What kind of mechanism is involved and under what kind of hormonal control is resorption from the bladder achieved? All these questions remain unanswered.

The non-burrowing (hylid, rhacophorid, hyperoliid) frogs do not generally survive prolonged dehydration, nor are they capable of storing water in their bladder in quantities that would enable them to survive longer periods. However, even a hylid as small as *Litoria rubella* was capable of surviving for 8 h in dry air at high temperatures (40 °C) (Warburg 1965). Size did not seem to affect dehydration tolerance (Newman and Dunham 1994) *Salamandra salamanadra* is also capable of surviving water loss for prolonged periods (Degani 1981a,b, 1994).

In addition, some fossorial species are known to form cocoons (Table 3.1). This form of protection has been seen among non-xeric amphibians (e.g. *Siren* sp., see Etheridge 1990a,b). Among anurans, several species of toads (*Bufo arenarum* and others) are capable of storing dilute urine in their bladder and peritoneal cavity (Schmajuk and Segura 1982).

5.1.5
Water Uptake

Water can be absorbed through amphibian skin. Of special significance is the skin on the ventral surface, in particular around the pelvic region, which plays an important role in water absorption from moist surfaces (Christensen 1974). Dehydration in *Bufo woodhousei* (and *B. marinus*) reduces this perfusion, and when water comes into contact with the pelvic patch, the result is increased water uptake (Malvin et al. 1992). In *Scaphiopus couchi* and *Bufo cognatus*, water uptake through the skin is stimulated by β-adrenergic agents (Hillyard 1979; Yokota and Hillman 1984, respectively).

Water uptake after dehydration is a remarkably rapid process (Table 5.5; Fig. 5.7). Most of the water is absorbed during the first 2 h (Warburg 1965,

Table 5.6. Bladder urine content

Species	Body weight (g)	Bladder urine content (% body weight)	Reference
Anura			
Bufonidae			
Bufo cognatus	23.2–81.9	19–31	Ruibal (1962)
B. cognatus	15.4	37.7	Schmid (1969)
B. americanus	22.2	23.8	Schmid (1969)
B. hemiophrys	18.4	28.4	Schmid (1969)
Hylidae			
Cyclorana platycephalus	5.9–13.2	57	van Beurden (1984)
Pelobatidae			
Scaphiopus couchi	22.1–29.2	20 ml/1100 g wt	McClanahan (1967, 1972)
Ranidae			
Pyxicephalus adspersus	10–100	14.1–32	Loveridge and Withers (1981)

1967). This physiological process is apparently not directly affected by temperature (Cloudsley-Thompson 1967; Figs. 5.8, 5.9). There is, however, evidence to suggest a relationship between the degree of water loss and the rate of water uptake. The greater the degree of desiccation, the greater the amount of water absorbed (Brekke et al. 1991). Thus, in *Neobatrachus* spp., the degree of water loss affects the rate of water uptake (Bentley et al. 1958). On the other hand, no such relationship has been seen in *Heleioporus* spp. This agrees with the observations by Thorson (1955), who found no relationship between the rate of water uptake following dehydration and the degree of terrestrial adaptation. Water uptake following dehydration of 25% ranged between 44 and 60 mg cm^{-2}h^{-1} in *Heleioporus* spp. and between 33 and 99 mg cm^{-2}h^{-1} in several *Neobatrachus* spp. (Bentley et al. 1958).

Heleioporus eyrei has been shown to be capable of replenishing lost water when burrowing in moist soil (Lee 1968). The exchange of water between salamanders and the soil is a function of the moisture tension of the soil or its degree of moistness (Spight 1967b). *Salamandra salamandra* is capable of replenishing its moisture from the moist soil (Warburg and Degani 1979). The fossorial anurans *Scaphiopus couchi*, *S. hammondi* and *Bufo cognatus* are capable of absorbing moisture from soil at a moisture tension of 2.5 atm (Walker and Whitford 1970).

The rate of water uptake by arboreal hylids (*Litoria caerulea, L. rybella, L. moorei* and *L. latopalmata*) and of some fossorial frogs (*Cyclorana platycephalus* and *Notaden nichollsi*) was examined by Main and Bentley (1964). The rate of water uptake of the burrowing leptodactylids (*Neobatrachus, Limnodynastes, Notaden*) was greater than that in the non-burrowing hylids (Warburg 1965, 1967; Heatwole et al. 1971). However, the fossorial *Pelobates*

Fig. 5.7. Water uptake in *Salamandra* during 4h

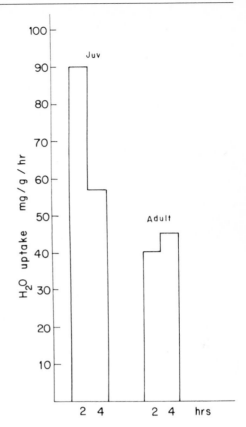

syriacus did not show a high rate of water uptake when compared with either *Bufo viridis* or *Hyla arborea* (Warburg 1971a).

The rate of water uptake by (*Chiromantis petersi* is 75% body weight/h (Drewes et al. 1977), and in *Hyperolius viridiflavus* it is over 69% of body weight (Geise and Linsenmair 1986). There is no difference in water uptake between frogs in the dry and wet seasons (Geise and Linsenmair 1988).

Among urodeles, the rate to water uptake of *Triturus vittatus* was considerably greater than that of *Salamandra salamandra* (Warburg 1917b). The heavier salamanders (*Ambystoma tigrinum*) gained more water following dehydration than the smaller ones (Gehlbach et al. 1969). *Batrachoseps* spp. more than doubled the percentage of water uptake during rehydration (Jones and Hillman 1978). In *Limnodynastes salmini*, the rate of rehydration was twice as high when in groups of 5 animals (Johnson 1969b). A similar phenomenon has been observed in the rehydration response of groups of juvenile *Pelobates syriacus* (Warburg 1917a).

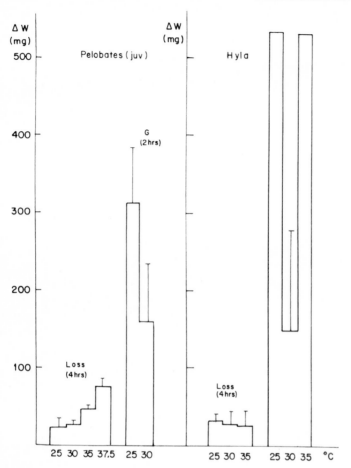

Fig. 5.8. Water loss over 4 h and water uptake over 2 h in *Pelobates* juveniles and in *Hyla savignyi*

5.2
Body Fluid Osmolality and Ion Concentration

This subject has been discussed by a number of authors (Alvarado 1979; Maloiy 1979; Shoemaker 1987, 1988; Shoemaker et al. 1992). Both plasma osmolality and ion concentration are dependent on the physiological state of an animal. Plasma osmolality ranged between 200 and 250 mOsm kg^{-1} (Alvarado 1979). Some of the variability seen in Table 5.7 can be attributed to this factor. Aquatic amphibians have a less concentrated plasma and therefore lower osmolality than terrestrial ones do (Schmid 1965b). The plasma osmolality in tadpoles of *Scaphiopus hammondi* increased towards metamorphic climax from less than 200 mOsm/l to over 300 mOsm l^{-1} (Funkhouser 1977). In those inhabiting xeric locations, the plasma osmolality ranged between 170

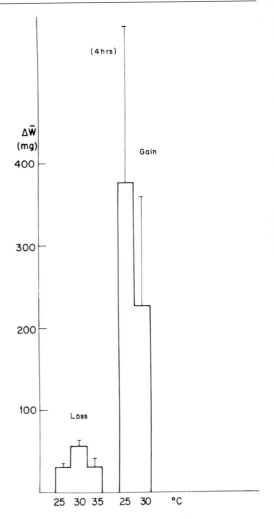

Fig. 5.9. Water loss and water uptake in the terrestrial phase of *Triturus vittatus*

and 350 mOsm, whereas the sodium concentration ranged between 45 and 120 mEq l^{-1}. In pelobatids the range was similar (Fig. 5.10).

Sodium and chloride fluxes are to some extent independent of each other in *Bufo arunco* (Salibian et al. 1968, 1971) and in *Ambystoma tigrinum*. Electrolyte exchange in anuran tadpoles and urodele larvae is across the gills, whereas in adults it occurs through the skin (Stiffler 1994). As the potassium concentration is generally low, urea must account for a comparatively high osmolality. The percentage of urea increased during metamorphosis. On the other hand, sodium levels are lower in *Pelobates* juveniles than adults (Degani and Nevo 1986).

When animals are dehydrated, both the osmolality and the ion concentration increase markedly (Lee 1968; Degani and Warburg 1984; Degani et al.

Table 5.7. Plasma osmolality and ion concentration

Species	Osmolality (mOsm)	Ions (mEq/l)			Reference
		Na^+	K^+	Cl^-	
Anura					
Bufonidae					
Bufo boreas	156	64	4	61	Balinsky (1981)
B. marinus	209	127	4		Balinsky (1981)
B. viridis	356	113	6.5	111	Katz et al. (1986)
B. viridis	348	120.2	3.6	99.3	Schoffeniels and Tercafs (1965/1966)
B. viridis		112	5.8	89	Hoffman and Katz (1991)
B. viridis	330	120		100	Degani et al. (1984)
Hylidae					
Hyla pulchella		108	4.3	77	Shoemaker and McClanahan (1975)
Agalychnis annae		115	5.8	86	Shoemaker and McClanahan (1975)
Pachymedusa danicolor		108	5.5	87	Shoemaker and McClanahan (1975)
Phyllomedusa sauvagei		114	2.7	82	Shoemaker and McClanahan (1975)
P. pailona		95	4.6	71	Shoemaker and McClanahan (1975)
P. iherengi		121	4.7	72	Shoemaker and McClanahan (1975)
Ascaphus sp	172	106		81	Mullen and Alvarado (1976)
Pelobatidae					
Pelobates syriacus (juvs)		46.5	84.5	21.8	Degani and Nevo (1986)
Scaphiopus couchi	600				Bentley (1966a)
Rhacophoridae					
Chiromantis petersi	210	78.3	4.1	67	Drewes et al. (1977)
Urodela					
Salamandridae					
Salamandra salamandra	210–242	102–126	3.2–3.8	81–100	Degani (1981b)
Ambystomatidae					
Ambystoma tigrinum	262.2	107.1	4.8	115.1	Romspert and McClanahan (1981)
Ambystoma tigrinum	300.0				Delson and Whitford (1973b)
A. tigrinum (larvae)	220	103	0.9	74	Gasser and Miller (1986)
Plethodontidae					
Batrachosepus relictus		107.0		94	Licht et al. (1975)
B. attenuatus	310	107.8			Licht et al. (1975)

1983). Thus, *Heleioporus eyrei* showed an increased plasma sodium concentration following water loss (Lee 1968), and *Scaphiopus couchi, Bufo cognatus* and *Bufo borea* more than doubled their plasma sodium concentration during dehydration (Hillman 1980). The sodium concentration in dehydrated *S. couchi* reached $330\,mEq\,l^{-1}$. In *Hyla arborea*, plasma osmolality increased from 200 to $510\,mOsm\,l^{-1}$ when dehydrated, and the plasma sodium concentration increased more than the potassium concentration (Degani and Warburg 1984). The opposite was true in the muscles as the potassium concentration increased more than the sodium (Degani and Warburg 1984). Muscles of both *Bufo cognatus* and *Scaphiopus couchi* could tolerate a high osmotic concentration

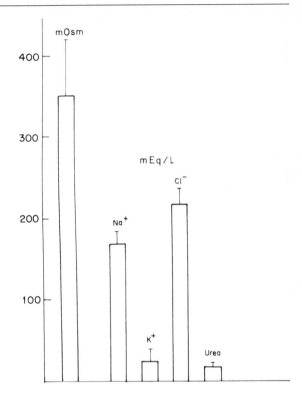

Fig. 5.10. Osmolality and ion concentration in *Pelobates* juveniles

(McClanahan 1964). *S. salamandra*, when dehydrated to 80% of its body weight, more than doubled its plasma osmolality and the plasma sodium concentration (Degani 1981b).

Similarly, in *Pelobates syriacus*, the sodium concentration increased markedly in both plasma and muscle following dehydration (Degani et al. 1983). However, it was not possible to acclimate it to osmolalities higher than 450 mOsm (Shpun et al. 1992). Both *Hyla arborea* and *B. viridis* could be acclimated to 500 mOsm l^{-1} (Shpun et al. 1992). Salt loading in *Batrachoseps attenuatus* and *B. relictus* leads to elevated plasma osmolality up to 600 mOsm l^{-1} (Licht et al. 1975).

There was a difference between cocooned and non-cocooned *Phyxicephalus adspersus*, as dehydrated cocooned frogs had a higher plasma Na$^+$ and osmolality than had the hydrated, non-cocooned frogs (Loveridge and Withers 1981). In the urodele *Batrachoseps* spp., the plasma osmolality reached 670 mOsm when salt-loaded (Jones and Hillman 1978).

In *Bufo viridis*, plasma osmolality has been recorded as 247 and 270 mOsm l^{-1} (Gordon 1962, Katz 1989). In the natural population, however, it proved to oscillate slightly throughout the year around 300 mOsm kg^{-1} (Katz et al. 1986). In a population living on saline soil, the plasma sodium concentration was $274\,mEql^{-1}$ (Khlebovich and Velikanov 1982). When acclimated to 40% seawater both plasma, sodium, chloride and urea increased markedly in *Bufo marinus* (Liggins and Grigg 1985). In the fossorial pelobatids *Scaphiopus couchi* and *Pelobates syriacus*, the plasma concentration fluctuates (Mc-Clanahan 1972; Degani et al. 1983). When dehydrated or buried in the soil for 3 months, the plasma osmolality of *P. syriacus* was 375 and $415\,mOsm\,kg^{-1}$, respectively (Degani et al. 1983; Fig. 5.10).

During dehydration, both the muscle and plasma urea concentrations of *B. viridis* increased many times, reaching $611\,mmol^{-1}$, while osmolality reached $1000\,mmoll^{-1}$ (Degani et al. 1984; Hoffman et al. 1990; Hoffman and Katz 1991). In a natural population of *B viridis*, the urea concentration in the plasma fluctuated (Katz et al. 1986). The frogs could be acclimated to urea with high osmolality ($500\,mOsm/kg^{-1}$; Katz et al. 1984). It appears that the increased rate of urea production is short-termed (Hoffman and Katz 1994).

In *S. salamandra* from a xeric habitat, the blood plasma concentration is higher than in individuals from mesic habitats (Degani 1981b, 1982a). In both populations, the plasma osmolality was highest ($340\,mOsm\,kg^{-1}$) towards the end of the summer compared with the winter value of 220–$270\,mOsm/kg$ (Degani 1985b; Katz 1989). Under experimental conditions, e.g. when in slowly drying soil for 2–3 months, *Bufo viridis* has shown increased plasma osmolality up to $1034\,mOsml^{-1}$, coupled with increased concentrations of sodium (Katz and Hoffman 1990).

In conclusion, it is difficult to establish that xeric-inhabiting amphibians differ from mesic ones in their water balance. It is likely that the differences are quantitative rather than qualitative ones. That is, they can survive longer without water and perhaps replenish it without difficulty by taking up water more quickly and in larger quantities.

5.3
Nitrogen Balance

5.3.1
Ureotelic Amphibians and Urea Retention

Aquatic amphibians have a low level of plasma urea and a high ammonia content in their urine (Schmid 1968). Most of the ammonia is formed in the kidneys, whereas some may be excreted through the skin (90% in *Necturus*, see Fanelli and Goldstein 1964; less than 15% in *Xenopus*, see Balinsky and Baldwin 1961). Newts excrete a higher percentage of ammonia when in water during the breeding season (see Table VI in Balinsky 1970). On the other hand, most terrestrial amphibians, including the fossorial and species inhabiting xeric environments studied so far, are ureotelic (McClanahan 1975). This is due to the non-toxic nature of urea, which can be accumulated during re-stricted urine flow (Balinsky et al. 1961, 1967). Under dry conditions, *Xenopus* switches from ammonotelism to ureotelism (Balinsky 1970). Thus, urea is the main nitrogen excretion product (over 90%) in *Scaphiopus couchi* and *S.*

multiplicatus (Jones 1980a,b), *Bufo americanus and B. hemiophrys* (Schmid 1968), *B. arenarum* (Salibian and Fichera 1984), *B. arunco* (Rovedatti et. al. 1988), *B. viridis* (Lee et al. 1982) and *Salamandra salamandra* (Cragg et al. 1961). Some of these data are given in Table 5.8.

Embryos of *Pseudophryne corroboree* do not excrete nitrogen immediately prior to hatching (Domm and Janssens 1971). After hatching, the tadpoles excrete ammonia. Most amphibians begin as ammonotelic tadpoles or larvae switching to ureotelism upon metamorphosis. Although *Ambystoma* larvae do excrete ammonia, there is evidence that they are also ureotelic (Dietz et al. 1967; Stiffler et al. 1980). The urea concentration in the nitrogen end products increases towards metamorphic climax in the aquatic *Xenopus*, peaking in the juveniles (Schultheiss and Hanke 1978). In tadpoles of the leptodactylid *Caudiverbera caudiverbera*, total urea excretion increased about 30% towards metamorphic climax (Zamorano et al. 1988). Ureotelism enables tadpoles of *Scaphiopus* spp. to survive for short periods out of water, even before they have completed metamorphosis (Jones 1980a). They are also capable of accumulating urea in their body fluids (Jones 1980a). Some tadpoles of land-nesting leptodactylids (*Leptodactylus bufonius*) produce urea (Shoemaker and McClanahan 1973). Similarly, intrauterine larvae of *Salamandra salamandra* are ureotelic, and the total nitrogen content of the uterine fluid in female salamanders during gestation contained 20% more urea than that of non-pregnant females (Schindelmeiser and Greven 1981).

Although the major component of nitrogen excretion in post-metamorphic amphibians is generally urea (*Hyperolius nasutus*, see Withers et al. 1982a), some adult leptodactylids (*Caudiverbera caudiverbera*) excrete only about half of their nitrogen end products in this form (Espina et al. 1980). Others (*Rana*) can excrete ammonia through the skin (Frazier and Vanatta 1980).

Several amphibian species are capable of synthesising and accumulating large amounts of urea in their body fluids and tissues (Jungreis 1976; Balinsky 1981; Jones 1982). *Chiromantis petersi* accumulates urea at a rare of $58.5\,mg\,N\,kg^{-1}day^{-1}$ (Drewes et al. 1977). Thus, elevated serum and tissue (muscle, liver) urea levels and increased urea cycle enzymes are a response to osmotically stressful conditions during dehydration or salt-loading (in *Xenopus*, Balinsky et al. 1961; in *Pelobates syriacus*, Degani et al. 1983; in *B. viridis*, Lee et al. 1982; Katz 1973; Degani et al. 1984; Degani 1985a; Hoffman and Katz 1991; Shpun et al. 1992; in *B. arenarum*, Boernke 1974; in *Hyperolius viridiflavus taeniatus*, Schmuck et al. 1988). In *B. viridis* this leads to increased plasma osmolality up to 1000 mOsm (Shpun et al. 1992).

Scaphiopus couchi accumulates high concentrations of urea during dormancy (McClanahan 1967). The maximum concentration of urea in starving *Bufo calamita* was greater in winter (650 mM) than in summer (410 mM; Sinsch et al. 1992). Within 42 days the amount of urea doubles (Whitford 1969). Kidney arginase levels are apparently affected by stressful environmental conditions, whereas liver arginase levels are not (Boernke 1973, 1974). On the other hand, diet affects the urinary nitrogen end products of *B. arenarum* (Castane et al. 1990).

Table 5.8. Nitrogen excretion in $mg\,N\,kg^{-1}\,day^{-1}$ unless stated otherwise

Species	Ammonia	Urea	Urate	Reference
Anura				
Bufonidae				
B. arunco	52.7	1148.1		Rovedatti et al. (1988)
B. viridis		35–36.1 mM l⁻¹		Gordon (1962)
B. viridis		20 mM		Schoffeniels and Tercafs (1965/1966)
B. viridis		33 mEq		Balinsky (1981)
B. boreas		51		Balinsky (1981)
B. marinus		36		Balinsky (1981)
Bufo woodhousei	45.7 µg g⁻¹ day⁻¹		153	Jones (1980a,b)
B. paracnemis		24.5 mM l⁻¹		Schoffeniels and Tercafs (1965/1966)
B. spinulosus		39		Schoffeniels and Tercafs (1965/1966)
B. arenarum		34.8		Schoffeniels and Tercafs (1965/1966)
B. arenarum	43.7 µg ml⁻¹	1186.5	1.3	Rovedatti et al. (1988)
B. fernandezae	34.3	822.9	1.0	Rovedatti et al. (1988)
Hylidae				
Hyla regilla	2.3	90.2	0.09	Withers et al. (1982a)
Phyllomedusa hypochondrialis	15.5	256	70.2	Shoemaker and McClanahan (1975)
			24	Cei (1980)
P. pailona	12.3	11.3	106	Shoemaker and McClanahan (1975)
			45	Cei (1980)
P. sauvagei	18.3 kg day⁻¹	39.6	212	Shoemaker and McClanahan (1975)
			90	Blaylock et al. (1976)
P. sauvagei	48	52		Shoemaker and McClanahan (1982)
P. iherengi	5.8	18.4	110	Shoemaker and McClanahan (1975)
Pachymedusa danicolor	2.5	5.5	7.6	Shoemaker and McClanahan (1975)
Agalychnis annae	13.7	113	16.2	Shoemaker and McClanahan (1975)

Hyperolidae				
Hyperolius nasutus	5.4 mM	71.3	0.02	Withers et al. (1982a)
H. viridiflavus taeniatus		124–433 mM		Schmuck and Linsenmair (1988)
Leptodactylidae				
Leptodactylus ocellatus	105.8 µg ml^{-1}	1122	0.6	Rovedatti et al. (1988)
L. bufonius		87.9		Cei (1980)
Pelobatidae				
Scaphiopus couchi	38.4 µg g^{-1} day^{-1}	62.2 µg g^{-1} day^{-1}	141–161	Jones (1980a,b)
S. multiplicatus	22.7		94–108	Jones (1980a,b)
Rhacophoridae				
Chiromantis xerampelina			190 mg 100 g^{-1}	Loveridge (1970)
C. petersi		4.4	149	Drewes et al. (1977)
Urodela				
Ambystomatidae				
Ambystoma tigrinum	0.1	15.3 mM l^{-1}		Romspert and McClanahan (1981)
A. tigrinum		0.9 mM		Gasser and Miller (1986)
A. tigrinum (larva)		41.9 mM		Gasser and Miller (1986); Romspert and McClanahan (1981)
A. mexicanum	74.3 µg g^{-1} day^{-1}	49.4		Cragg et al. (1961)
Salamandridae				
Salamandra salamandra	3.4 µg g^{-1}	62.7		Cragg et al. (1961)
S. salamandra		17		Degani et al. (1984)
S. salamandra	6.1 mM l^{-1}	15		Schindelmeiser and Greven (1981)
S. salamandra	84 mM l^{-1}	329 µM l^{-1}		Schindelmeiser and Greven (1981)
Plethodontidae				
Batrachoseps attenuatus	48 mEq			Balinsky (1981)
B. major	218			Balinsky (1981)

5.3.2
Uricotelic Amphibians

Uricotelism is the most significant water-conserving mechanism that allowed the archeosaurs to colonize arid habitats during the Triassic (Campbell et al. 1987; Wright 1995).

Recently, it has been found that some arboreal frogs inhabiting xeric habitats are capable of producing either urates or uric acid (Table 5.8). These frogs (rhacophorids and leptodacylids) remain exposed to the sun and wind even during the dry season (McClanahan 1975). Thus, the rhacophorid *Chiromantes xerampelina* produces 130–225 mg/100 g or 61–75% of its total nitrogen end products as urates (Loveridge 1970). It contains levels of the enzymes of uric acid biosynthesis which are five times higher than normal in *Bufo regularis* (Balinsky et al. 1976). On the other hand, *Chiromantes petersi* produces most of its nitrogen excretion products (97%) as uric acid (Drewes et al. 1977). Some phyllomedusine leptodactylids (*Phyllomedusa sauvagei, P. pailona, P. ihrenegi, P. hypochondrialis*) produce 80–230 mg/100 g uric acid, which is 80–95% of their total nitrogen products (Shoemaker and McClanahan 1975). However, when in water, *Phyllomedusa sauvagei* produced only 57% uric acid rather than 88% when on land (Shoemaker and Bickler 1979). In tadpoles of *P. sauvagei*, xanthine dehydrogenase is present in the liver, prior to the onset of uricotely at the metamorphic climax (Shoemaker and McClanahan 1982). Upon metamorphosis, xanthine oxidase is present at high levels, whereas uricase occurs at low levels in the liver and kidney of uricotelic phyllomedusine frogs (McClanahan 1975).

This phenomenon is also known in fossorial bufonids (*Bufo arenarum,* see Salibian and Fichera 1984), but not in pelobatids or hylids. Moreover, other frogs that remain exposed during the long dry summer months are not uricotelic (*Hyperolius nasutus*, Withers et al. 1982a,b; *H. viridiflavus taeniatus*, Schmuck and Linsenmair 1988; Schmuck et al. 1988).

5.3.3
Other Forms of Nitrogen Excretion

Purines, guanine, hypoxanthine and adenine occur in the form of white crystals in special cells known as the guanophores (Stackhouse 1966). In the hylids, large amounts of pteridines usually occur in the skin (Balinsky 1970). In the xeric-adapted frog *Hyperlius viridiflavus taeniatus*, deposits of guanine and hypoxanthine within the iridiophore crystals account for an increase in the dermal purine content. These deposits are harmless and can remain there indefinitely. That is how the frog deals with its nitrogen end products (Geise and Linsenmair 1986; Schmuck and Linsenmair 1988; Schmuck et al. 1988). Increased deposits of guanine have been found in light-coloured larvae of *Ambystoma tigrinum* inhabiting turbid water (Fernandez 1988).

5.4
Endocrine Control of Water, Electrolyte and Nitrogen Balance

Amphibians are the only vertebrates capable of living both in water and on dry land. Most terrestrial amphibians spend less than 5% of their lives in water (during breeding and during their metamorphic cycle). For most of their life, they are exposed to terrestrial conditions, sometimes even extreme ones. This mode of life requires elaborate mechanisms for maintaining water, electrolyte and nitrogen balance under vastly different environmental conditions. Thus, diuresis and ammonotelism are effective only when an amphibian is aquatic, whereas terrestrial life requires antidiuresis, ureotelism or even uricotelism.

Table 5.9. Amphibian species from xeric habitats in which the endocrine control of water, electrolytes or nitrogen balance was studied

Family/species	Hormones	Reference
Anura		
Bufonidae		
Bufo viridis	OXY, AVT, PRL	Warburg (1971a); Goldenberg and Warburg (1976, 1977a,b); Mack and Hanke (1977a,b)
B. arenarum	OXY	Uranga and Sawyer (1960); Uranga (1973)
B. marinus	AVT, OXY	Sawyer (1957); Jackson and Henderson (1976); Han et al. (1978)
B. boreas	AVT	Kent and McClanahan (1980)
Leptodactylidae		
Neobatrachus spp.	Pitocin	Bentley et al. (1958)
Pelobatidae		
Pelobates cultripes	AVT	Bentley and Heller (1965)
P. syriacus	OXY, AVT	Warburg (1971a)
Scaphiopus couchi	AVT, AVP	Hillyard (1976a,b)
Urodela		
Salamandridae		
Salamandra salamandra	OXY, AVT, PRL	Bentley and Heller (1964, 1965); Platt and Christopher (1977); Warburg (1971b); Warburg and Goldenberg (1978a,b); Wittouck (1972, 1975a)
Ambystomatidae		
Ambystoma tigrinum	AVT, MT, PRL, Aldosterone, Corticosterone	Alvarado and Johnson (1965, 1966); Bentley and Heller (1964); Stiffler (1981); Stiffler et al. (1982, 1984, 1986); Platt and Christopher (1977); Schultheiss (1977); Heney and Stiffler (1983); Jorgensen et al. (1947); Ireland (1973); Carr and Norris (1988); Norris et al. (1973); Wittouck (1972, 1975a,b); Brown et al. (1988)
A. gracile	PRL, corticosterone	Brewer et al. (1980)
A. mexicanum	Thyroid	Schultheiss (1977)
Triturus vittatus	OXY, AVT, PRL	Warburg (1971b); Warburg and Goldenberg (1978b); Stiffler (1981); Stiffler et al. (1984)

The neurohypophysial hormones regulating these functions must be secreted and circulating in the blood whenever required. On the other hand, each of these hormones must not function as a regulating agent continuously, but only when necessary. Consequently, there must be some readiness of the specific target organs (skin, kidney, bladder) enabling them to respond to hormonal action only during definite periods.

The following is an account of present knowledge regarding endocrine control of water, ion and nitrogen balance in amphibians inhabiting xeric environments (Table 5.9).

5.4.1
Endocrine Control

A large number of reviews on endocrinological control of water and nitrogen balance in Amphibia are available: Jorgensen (1950, 1993a); Sawyer and Sawyer (1952); Sawyer (1956); Heller and Bentley (1965); Heller (1965, 1970); Bentley (1969, 1987); Warburg (1972, 1988, 1995); Scheer et al. (1974); Bern (1975); Pang (1977); Sawyer et al. (1978); Alvarado (1979); Loretz and Bern (1982); Goldenberg and Warburg (1983); Hanke (1985); Hirano et al. (1987); Sawyer and Pang (1987); Brown and Brown (1987); Warburg and Rosenberg (1990); Olivereau et al. (1990); Hourdry (1993); Hanke and Kloas (1994).

Certain neurohypophysial hormones regulate water balance in amphibians. Their injection leads to a weight increase due to water uptake (Belehradek and Huxley 1928). Hypophysectomy produces a decrease in urine flow and GFR in larval *Ambystoma tigrinum* (Stiffler et al. 1984), as well as an anti-diuretic reaction of the bladder (Kerstetter and Kirschner 1971). There is some variability in the amount of hypothalamic neurosecretion (as judged by the phenology of the neurosecretory cells in *Scaphiopus hammondi* when exposed to different ambient conditions; Jasinski and Gorbman 1967).

5.4.1.1
The Neurohypophysial Hormones

There are a number of reviews on this subject (Sawyer and Pang 1975, 1987; Goldenberg and Warburg 1983; Warburg 1995).

Arginine Vasotocin – Synthesis of Arginine Vasotocin (AVT) in the Hypophysis, and Its Circulation in the Plasma. AVT is found in amphibians, located in the neurohypophysis in the median eminence and the neural lobe (Sawyer 1972; Sawyer and Pang 1975; Pang et al. 1983). In tadpoles the amounts are only a quarter of those found in adult frogs (Bentley and Greenwald 1970). The AVT concentration rises after dehydration due to a decrease in blood volume, and not because of increased osmolality (Nouwen and Kuhn 1983, 1985; Jorgensen 1993a). Dehydration also affects the levels of AVT in the ventral preoptic brain area (Zoeller and Moore 1986). In *Scaphiopus hamondi*, dehydration caused a decrease of hypothalamic neurosecretory granules (Jasinsky and Gorbman 1967). There is evidence of a seasonal effect on AVT levels which appears to be higher during the breeding season in the spring (Zoeller and Moore 1988), perhaps because of the breeding activity of the male. Little is known about AVT levels in amphibians from xeric habitats. Recently, however, Chauvet et al. (1993) identified AVT in *Bufo regularis*. Finally, is there a change in AVT levels when the tadpoles or larvae metamorphose and emerge onto land and, possibly, when urodele (newts) return to water to breed?

AVT Effect on Water Balance. AVT affects amphibians (especially Hylidae and Pelobatidae) by causing glomerular antidiuresis due to reduced GFR and consequently a rise in body water (Heller and Bentley 1965). It also leads to increased sodium uptake through the skin (Sawyer et al. 1978; Pang et al. 1983). Recently, Jorgensen (1991) cast some doubts on the role that AVT plays in the water economy of amphibians.

In urodele larvae, AVT effected a singificant increase in water uptake towards metamorphic climax (*Ambystoma tigrinum*, Alvarado and Johnson 1966). A similar effect has been seen in metamorphosing *Salamandra* larvae (Warburg and Goldenberg 1978a). In *Pelobates syriacus* and *Bufo viridis*, the water uptake following treatment with AVT was considerably lower than that following treatment with oxytocine (Warburg 1971a; Goldenberg and Warburg 1977b). A similar pattern has been observed in *S. salamandra* and *Triturus vittatus* (aquatic phase). To our surprise, the terrestrial phase of the newt (*T. vittatus*) responded to AVT in much the same way as to oxytocin (Warburg 1971b). The effects of AVT on water balance have been reviewed in Sawyer (1972) and Sawyer and Pang (1987).

AVT Effect on Water Movement: Epidermis. Osmotic permeability through *Bufo marinus* skin was three times higher following treatment with AVT (Bentley 1969). The permeability of the ventral pelvic epidermis of *B. boreas* and *B. bufo* increased following treatment with AVT (Christensen 1975; Kent and McClanahan 1980). However, AVT reduced the permeability of *B. viridis* (Mack and Hanke 1977b) and had no effect on the passage of water across the epidermis of either *Scaphiopus couchi* or *Ambystoma tigrinum* (Bentley 1969; Hillyard 1975, 1976a; respectively).

Urinary Bladder. An increase of 14% in body weight was noted in *Salamandra* following treatment with AVT (Bentley and Heller 1965). This was not seen, however, when the bladder was ligated. In bladder preparations of both *Salmandra salamandra* and *Triturus vittatus*, the response to AVT was considerably greater than that to oxytocin (Warburg 1971b). In *Bufo marinus*, water transfer was considerably enhanced following treatment with vasotocin (Bentley 1969).

Kidney Function. The weight gain observed "in vivo" by larvae of *Ambystoma tigrinum* appears to be the result of an antidiuretic response due to reduced GFR (Alvarado and Johnson 1965; Stiffler et al. 1982). There is evidence of increased tubular reabsorption as a result of AVT treatment (Ewer 1951, 1952a,b; Sawyer and Schisgall 1956; Bentley 1958, 1974) and an effect on the number of functional glomeruli in *Bufo marinus* (Jackson and Henderson 1976). In the spadefoot toad *Scaphiopus couchi*, evidence was found for reduced GFR after treatment with AVT (Hillyard 1975). There was also a lower rate of urine excretion in *Bufo marinus* (Sawyer 1957).

AVT Effect on Ion Balance. AVT caused sodium uptake and accumulation in larval *Ambystoma* (Jorgensen et al. 1947; Alvarado and Kirschner 1963; Alvarado and Johnson 1965). On the other hand, according to Bentley and Baldwin (1980), it did not affect sodium influx in *Ambystoma tigrinum* larvae (see also Kerstetter and Kirschner 1917). No effect was noticeable in the short circuit current (SCC) of *Ambystoma* larvae (Bentley and Heller 1964), whereas AVT caused a marked rise in SCC across the skin of *Bufo marinus* and *Scaphiopus couchi* (Bentley 1969; Hillyard 1976a). It also caused a decrease in sodium and plasma concentration in *Bufo viridis* (Degani 1984a).

Arginine Vasopressin (AVP). In *Bufo marinus*, AVP causes sodium and water transfer across the bladder (Bentley 1967a,b). When treated with vasopressin *Bufo melanostictus*, *B. boreas* and *B. punctatus* showed an increase in water uptake across the pelvic skin (Baldwin 1974). On the other hand, no effect was observed in *Scaphiopus couchi* (Hillyard 1976a,b) although AVP causes an elevation of SCC in the skin of *B. marinus*.

Mesotocin (MT). MT is present in amphibians and causes both diuresis and increased GFR in *Ambystoma tigrinum* larvae (Stiffler 1981; Stiffler et al. 1984) and a net loss of Na^+ in *Bufo arunco* (Salibian 1977). It also produces a hydrosmotic effect on the bladder (Sawyer et al. 1978).

Oxytocin (OXY). Most experiments have been conducted with OXY, a hormone not occurring naturally in Amphibia, but easily available commercially. The animals responded to this hormone by gaining weight (accumulating water) or by increasing the permeability of either the skin or the urinary bladder to water.

OXY Effect on Water Balance. Larvae and post-metamorphic stages of *Bufo viridis*, *Pelobates syriacus* and *Salamandra salamandra* responded by taking water through their skin after only 1 h following treatment with OXY (Goldenberg and Warburg 1976; Warburg and Goldenberg 1978a). A very low response was described, however, in *Pelobates* tadpoles and in adult *Salamandra*. We noted the highest response in the juvenile stage. In *Bufo viridis*, the response was significant towards the metamorphic climax, reaching a peak at the juvenile stage (Goldenberg and Warburg 1976). In adult *Bufo viridis* and *Pelobates syriacus* the response to OXY was also significant, although less so than in juveniles (Warburg 1971a). OXY also caused cutaneous water uptake in *Bufo melanostictus* (Elliott 1968) and increased cutaneous water uptake in *B. marinus* (Bakker and Bradshaw 1977).

In *B. marinus*, osmotic water transport following treatment with OXY was greater in the winter (cold) months than in the summer (warm) months (Han et al. 1978). The premetamorphic larvae of *S. salamandra* also responded to OXY (Warburg and Goldenberg 1978b), and the response reached its peak upon metamorphosis. Nevertheless, it was clearly noticeable in juveniles, and to some extent in adults (Warburg 1971a). No response to OXY was noted in *Ambystoma* larvae (Stiffler et al. 1984). The terrestrial phase of the newt (*Triturus vittatus*) has, on the other hand, shown some response to OXY (Warburg 1971b). In the adult *B. arenarum*, OXY caused an increase in GFR (Uranga and Sawyer 1960), without producing any change in the number or size of the glomeruli (Uranga 1973).

Water Movement Through Isolated Bladder and Epidermis. OXY caused increased water movement though the isolated bladder of *Bufo viridis, Pelobates syriacus, Hyla arborea, Salamandra salamandra* and *Triturus vittatus* (Warburg 1971a,b). The isolated epidermis was likewise affected by OXY through increased water movement in adult *Pelobates syriacus, Salamandra salamandra* and *Triturus vittatus*, but no response was noted in *B. viridis* (Warburg 1971a,b).

Hydrin 2. This neurohypophysial hormone found only in some anuran amphibians was recently identified also in the toad *Bufo regularis* (Chauvet et al. 1993). It is involved in the endocrine control of osmoregulation by increasing skin and bladder permeability to water (Rouille et al. 1989). In *Bufo regularis* and *B. viridis*, the ratio of hydrin 2 to vasotocin is 2, whereas in mesic anurans it is between 0.8 and 1.4 (Michel et al. 1993).

5.4.1.2
The Adenohypophysial Hormones Prolactin (PRL)

The neotenic axolotl (*Ambystoma tigrinum*) contains more PRL than do either larvae or adults (Norris et al. 1973). PRL-producing cells were identified in the pituitary gland of *Hyla arborea* (Campantico et al. 1985; see Fig. 5.11).

PRL Effect on Water Balance. PRL causes a net uptake of water in amphibians (Brown and Brown 1987; Hirano et al. 1987). In *Bufo viridis*, the four-limbed tadpole gained weight when treated with PRL, responding more than the juvenile (Goldenberg and Warburg 1977a). In *Ambystoma tigrinum*, PRL also caused an increase in body water (Platt and Christopher 1977). In the salamander larva (*Salamandra salamandra*) PRL caused water retention in the 7-day-old larva, and more so in juvenile salamanders (Warburg and Goldenberg 1978b). On the other hand, adult *Bufo viridis* were affected more by PRL than

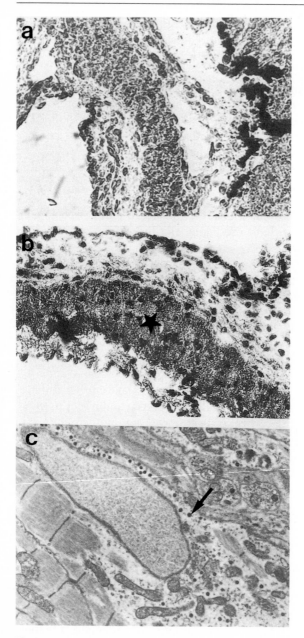

Fig. 5.11. Reaction to atrialnatriuretic peptide (ANP) in the heart of *Pelobates*. *Top* Section through the atrium of a legless *Pelobates* tadpole showing no reaction to BNF (brain natriuretic factor; ×160). Atrium of a two-legged tadpole shows strong positive reaction to treatment with ANP antibodies (*star* in **b**, ×160). Note the putative ANP granules (*arrow*) in an atrium cell of a juvenile toadlet (**c**, ×5000)

adult *Salamandra*. In *Bufo marinus*, PRL produced increased osmotic water flow through the isolated bladder (Snart and Dalton 1973; Debnam and Snart 1975).

PRL Effect on Electroloyte Balance. PRL also affects ion movement. Thus, it increased the ion-absorbing capacity of *Ambystoma* gills (Wittouck 1972), as well as sodium (and calcium) retention in the axolotl *Ambystoma maexicanum* (Wittouck 1972). It also increased sodium transport in *Ambystoma tigrinum* (Platt and Christopher 1977) and in *A. gracile* (Brewer et al. 1980), but no effect on sodium levels in *Ambystoma tigrinum* (Wittouck 1975a). SCC increased at metamorphic climax and as a result of treatment with PRL (Eddy and Allen 1979). This was also shown "in vitro" in the bladder of *Bufo marinus* (Snart and Dalton 1973; Hirano et al. 1987).

PRL Effect on Integument. PRL affects the amphibian integument in various ways (Dent 1975). It stimulates the secretory activity of the mucous glands by causing the release of granular secreation (Hoffman and Dent 1978). This increased mucus production has been observed in the axolotl *Ambystoma tigrinum* (Wittouck 1975b) and in *A. gracile* (Brewer et al. 1980). PRL also led to an increase in the number of mucous glands in *Triturus* (Vellano et al. 1970). It produced a regression of the cornified tubercles in the skin of newts (Dent 1975). It appears to affect the "water drive" or movment towards water (see Chap. 3) and the integumentary changes related to this phenomenon.

5.4.1.3
The Thyroid Hormones

Both T_3 and T_4 were located in the plasma of *Ambystoma tigrinum* larvae, where they remained until the metamorphic climax. The plasma T_3 concentration was higher than that of T_4 (Larras-Regard et al. 1981). Both circulation and seasonal variations in the activity of the thyroid are reported by Kuhn et al. (1985). Thus, both T_3 and T_4 show a seasonal cycle in *Rana perezi* which peaked in July. There are reports that thyroxin leads to a rise in urea excretion in the axolotl *Ambystoma mexicanum* (Schultheiss 1977).

5.4.1.4
Corticosteroids and the Interrenal Hormones

The adrenal gland in Amphibia is situated on the ventral side of the kidney. Its position and length vary in the different amphibian groups (Grassi Milano and Accordi 1983, 1991). Two main types were recognized: the anuran type which is more median and diffuse, and the urodelan type which is more aggregated and lateral in position (Grassi Milano and Accordi 1986). Thus, in *Pelobates* it is considerably longer than in either *Bufo* or *Hyla*. It consists of chromaffin cells intermingled with cortico-steroidogenic cells (Picheral 1970; Accordi and Grassi Milano 1990). These cells in the adrenal cortex are responsible for the secretion of the adrenocortical steroids, ACTH and angiotensin II. In addition, the adrenal medullary cells synthesize somatostatin, neurotensin and epinephrine. The secretion of adrenocortical steroids is stimulated by ACTH (Alvarado 1979). Aldosterone and corticosterone are the principal corticosteroids in amphibians (Bentley 1971; Scheer et al. 1974). They vary seasonally and cycle with the diurnal rhythm (Adler and Taylor 1981).

In *Bufo arenarum*, interrenal cells appear to be more active during summer (Jolivet-Jaudet and Ishi 1985). Corticosterone and aldosterone levels in the plasma increase towards metamorphic climax (Jaffe 1981; Dent 1988; Jolivet-Jaudet and Leloup-Hatey 1984). Plasma corticosterone levels were low in premetamorphic stages of *Ambystoma tigrinum*, increasing during metamorphosis and decreasing towards metamorphic climax (Carr and Norris 1988). Similarly, the aldosterone concentration in embryonic *Bufo arenarum* peaks towards metamorphic climax (Castane et al. 1987).

Corticosterone caused a decline in sodium efflux in *Ambystoma tigrinum* (Brown et al. 1986), whereas in the neotenous *A. gracile* it led to sodium retention (Brewer et al. 1980). Similarly, aldosterone produced sodium retention and potassium loss in *Triturus* (Socino and Ferreri 1965) and in *A. tigrinum* (Stiffler et al. 1986).

Towards metamorphic climax, the aldosterone content rose sharply in *Bufo arenarum* tadpoles (Castane et al. 1987). The colon, integument and urinary bladder are the main sites of action for aldosterone (not the kidney; Henderson and Kime 1987). It stimulated sodium transport through the urinary bladder of *Bufo marinus* (Crabbe 1961; Voute et al. 1972) and the skin of *Rana* (Voute et al. 1975), but did not affect the plasma sodium concentration in *B. marinus* (Middler et al. 1969). Apparently, in hylids the ventral skin responds more to aldosterone than the dorsal skin by a factor of 10–20 times (Yorio and Bentley 1977). In the newt, aldosterone caused sodium retention (Socino and Ferreri 1965), and it stimulated the reabsorption of sodium and the secretion of K^+ in *A. tigrinum* (Heney and Stiffler 1983; Stiffler et al. 1986). It also stimulated sodium influx through both the skin and gills of *Ambystoma tigrinum* larvae (Alvarado and Kirschner 1963, 1964; Alvarado 1979). It stimulated sodium transport and caused an elevated SCC in *Bufo marinus* and *B. viridis* (Porter 1971; Nagel and Katz 1991). Aldosterone, ACTH and corticosterone all produced an elevation of plasma urea concentration in *Ambystoma mexicanum* (Schultheiss 1973, 1977).

Angiotensin II (A II). II leads to increased aldosterone production (Nishimura 1987). This hormone is apparently produced by special cells in the ventral preoptic and hypothalamic regions of the brain (Propper et al. 1992). II has been shown to have an antidiuretic and antinatriuretic effect on the kidney tubules of *Bufo paracnemius* (Coviello 1969). It also affects water absorption by the urinary bladder (Tran et al. 1992). Treatment with this hormone causes an increase in water absorption behaviour and water uptake in *B. punctatus* (Hoff

Seckendorff and Hillyard 1991, 1993a,b). On the other hand, A II may not be directly involved in the regulation of water absorption through the skin of *B. arenarum* (Soria et al. 1987). The renin-angiotensin system does not affect the plasma sodium levels (Nolly and Fasciolo 1971a,b).

5.4.1.5
Parathyroid and Ultimobranchial Hormones Regulating Calcium Balance

Amphibians possess ultimobranchial bodies which secrete the calcaemic hormone calcitonin (CT: Coleman 1975; Robertson 1987). This hormone has a hypocalcaemic effect (Uchiyama 1980). It blocks the resorption of calcium from the skeleton and the endolymphatic system (Bentley 1984). Parathyroid hormone (PTH) has a hypercalcaemic effect in some amphibians such as *Rana pipiens* and *R. catesbeiana* (Cortelyou and McWhinnie 1967; Uchiyama and Pang 1981), but not in others including *Bufo marinus* (Bentley 1983). The calcium stores in the bone, endolymph and skin may be the sites of PTH action. Similarly, PRL increases the mobilization of $CaCO_3$ from the endolymphatic sacs into the plasma (Campantico et al. 1974). The calcium stores in the dermis and endolymphatic system are related to the water balance. The lymph system is highly developed in amphibians, consisting of pulsating organs (hearts) and vessels. It was recently shown that the rate of lymph flow is greatly reduced following dehydration (Jones et al. 1992). Upon rehydration, apparently most of the water taken up through the skin finds its way first into the lymphatic system and only then into the plasma (Wentzell et al. 1993). Concurrently with dehydration, calcium is mobilized from the endolymphatic sacs (Campantico et al. 1974).

5.4.1.6
Atrial Natriuretic Peptides (ANP)

Atrial natriuretic peptides (ANP) are secreted by special cells (types A,B,C) found in the heart, kidney and other organs. In the frog they are located in the atrium, ventricle, brain and central nervous system (Netchitailo et al. 1986, 1987, 1988; Fig, 5.11), as well as in the kidney (Kloas 1992). A high intensity of ANP antibodies was found in the bulbus arteriosus of *Pelobates syriacus* two-legged tadpoles (Gealekman 1996).

ANP plays an important role in maintaining the homeostatic balance of salt and water in the body fluids (Goetz 1988; Evans 1990). It is a diuretic and natriuretic hormone causing increased GFR and a decrease in aldosterone, catecholamine and hypophysial hormone production. In a way, it counteracts the antidiuretic effects of the neurohypophysial hormones and A. II (Kloas 1992). The ANP-secreting cells of amphibians show cyclical changes during the metamorphic cycle (Hirohama et al. 1989): in *Bufo*, they are apparently more numerous in tadpoles than during the terrestrial post-metamorphic phase. These is very little information about these peptides, and no experimental work has so far been carried out on them.

5.4.1.7
Conclusions

Two main hormones regulate water balance in amphibians: MT acting as a diuretic agent and AVT, an anti-diuretic hormone. In addition, PRL, aldosterone, corticosterone and ANP are also effective in regulating water and ion balance. The hormones affect the epidermal and bladder permeability to water and ions as well as that of the kidney through control of the GFR. The main question concerns the presence of the hormones in amphibians. To what extent are they always present? Are these hormones being synthesized at all times, and under what circumstances are they being stored or released? Would the target organs respond in the same way at all periods? What happens with the endocrine control when an amphibian returns to water to breed?

The problem is that under most circumstances an amphibian in its aquatic environment probably responds differently than when it is terrestrial. Only partial information concerning the hormonal control of water balance is available at the moment, and it is based on only a small number of amphibian species which are not necessarily adapted to xeric environments.

5.5
Thermal Balance

The thermal relations of amphibians can be investigated in various ways (Table 5.10). The ecological approach measures the temperature ranges at which individual animals can survive either as aquatic larvae or in the terrestrial post-metamorphic stages. One way of doing this is by following the changes in body temperature of amphibians in the field. The ambient temperature conditions prevailing in either aquatic or terrestrial habitats are studied at the same time. The behavioural approach investigates the thermal preferences of individual animals in either aquatic or terrestrial environments. Some amphibians retreat into rock crevices (hylids) or burrow into the ground (pelobatids, leptodacylids, bufonids), whereas others bask in the such in order to raise their temperature. Finally, physiological investigations deal with metabolic temperatures, thermal acclimation, thermoregulation by evaporative cooling, survival, critical maximal temperatures (CTM) and lethal temperatures in either the aquatic or terrestrial milieu (see reviews by Brattstrom 1970b, 1979; Spotila et al. 1992; Hutchison and Dupre 1992). Thermal acclimation plays a role in adjusting thermal preferences as well as in lethal temperatures and the CTM. The ability to adjust is characteristic of species with a wide geographic range (Warburg 1967; Brattstrom 1968).

5.5.1
Temperature Preferences

The thermal preferences of the aquatic larval stages differ from those of the terrestrial, post-metamorphic stages. Thus, tadpoles of *Bufo marinus* show

Table 5.10. Thermal conditions of selected amphibian species

Family/species	Tadpole/Larva		Adult		Region	Reference
	CTM	Field	CTM	Lethal		
Bufonidae						
Bufo marinus	44–45				Caribbean	Heatwole et al. (1968)
B. marinus		27.2			Australia	Johnson (1972a)
B. marinus	42		40		N. America	Krakauer (1970)
B. marinus			41		N. America	Stuart (1951)
B. marinus			41.8		Panama	Brattstrom (1968)
B. cognatus		23.2			Panama	Brattstrom (1963)
B. cognatus			39.7		Mexico	Brattstrom (1968)
B. cognatus			38		Mexico	Schmid (1965a)
B. carens			35.5		Africa	Balinksy (1969)
B. regularis			36		Africa	Balinsky (1969)
B. americanus		24.7			N. America	Brattstrom (1963)
B. americanus			38		N. America	Brown (1969)
B. americanus	37–42				N. America	Brattstrom (1963)
B. alvarius		21.5			N. America	Brattstrom (1968)
B. alvarius			38.7		Mexico	
B. punctatus	32.6				Mexico	Moore and Moore (1980)
B. viridis		37.2–39.1			Asia	Warburg (1965)
B. boreas		25–30			Mexico	Lillywhite et al. (1973)
Hylidae						
Hyla arborea			39.1		Asia	Warburg (1971a)
H. smithi			42.5		Mexico	Brattstrom (1968)
Litoria caerulea		27.8	41.6		Australia	Johnson (1970)
L. caerulea			39.4		Australia	Johnson (1970, 1971a)
L. rubella		39.2	40.4		Australia	Main (1968); Warburg (1972)
Cyclorana cryptotis			42		Australia	Tyler et al. (1982)
C. cultripes		39.2			Australia	Main (1968)
C. albogusttatus			40		Australia	Brattstrom (1970a)
Hyperolidae						
Hyperolius viridiflavus nitidulus			43–44		Africa	Geise and Linsenmair (1986, 1988)

Table 5.10 (Contd.)

Family/species	Tadpole/Larva		Adult		Region	Reference
	CTM	Field	CTM	Lethal		
Myobatrachidae						
Limnodynastes dorsalis			35		Australia	Brattstrom (1970a)
Notaden nichollsii			33.6		Australia	Main (1968)
Leptodactylidae						
Eleutherodactylus portoricensis			36–38		Caribbean	Heatwole et al. (1965)
Leptodactylus albilabrius	41.3				Caribbean	Heatwole et al. (1968)
Leptodactylus melanotus			39.8		Mexico	Brattstrom (1968)
Microhylidae						
Phrynomerus annectens	42				Africa	Channing (1976)
P. adspersus				36	Africa	Balinsky (1969)
P. natalensis				36	Africa	Balinsky (1969)
Pelobatidae						
Pelobates syriacus			36.7		Asia	Warburg (1965)
Scarphiopus couchi		23.5	37.5		N. America	Brown (1969)
S. hammondi			37.5		N. America	Brown (1969)
			40.3		N. America	Brattstrom (1968)
Ranidae						
Tomopterna dellalandei	42				Africa	Channing (1976)
Pixicephalus adspersus				36	Africa	Channing (1976)
			38.5		Africa	Balinsky (1969)
P. delalandei			36		Africa	Balinsky (1969)
Ambyostomatidae						
Ambystoma tigrinum		21.2–22.6	35.8–36.7			Brattstrom (1963)
A. tigrinum						Claussen (1977)
Salamandridae						
Salamandra salamandra infraimaculata	35–35.9		35.7–36.2			Degani (1982b)

a preference for 27–30 °C, depending on their developmental stage, physiological state, and their acclimation temperature (Floyd 1984; Malvin and Wood 1991). As a result, when dehydrated, their temperature preference was lower by 8.6 °C. In *Bufo americanus* tadpoles the preferred temperature range was between 27.5 and 37 °C, whereas in *Bufo boreas* it was 28–34 °C (Beiswenger 1978), and higher in *B. cognatus* (Sievert 1991). On the other hand, Lillywhite et al. (1973) found differences between the thermal preference of *B. boreas* tadpoles in the field and in the lab. The thermal preference in the field was between 23.5 and 26.2 °C, and in the lab between 26 and 27 °C (Carey 1978).

Adult *Bufo boreas* toads preferred 24.2 °C under field conditions (Smits 1984). The maximum temperature recorded in *B. mauritanicus* in the Sahara desert of Morocco was 33 °C (Meek 1983). Adult anurans studied by Strubing (1954) have shown a thermal preference for 29.8 °C (*Hyla arborea*), 33 °C (*Bufo viridis*) and 33.1 °C (*Hyperolius horstocki*). In *Hyla arenicolor* the body temperature in summer was 30.7 °C when the exposed rock temperature was 45 °C (Snyder and Hammerson 1993). Likewise, thermal preference by *Bufo marinus* was higher in spring than in fall (Mullens and Hutchison 1992). Fully hydrated *Chiromantis xerampelina* "preferred" a temperature range between 35 and 38 °C, whereas *Phyllomedusa sauvagei* did not show any thermal preference between 31 and 41 °C (Shoemaker et al. 1989).

Ambystoma tigrinum larvae preferred a temperature range between 16 and 31 °C (Lucas and Reynolds 1967), whereas juvenile *A. maculatum* preferred 32 °C (Pough and Wilson 1970). *Salamandra salamandra* larvae from moist habitats preferred somewhat lower temperature ranges than those preferred by larvae from more xeric habitats (Degani 1984a). No diurnal rhythm in temperature selection was observed (Mullens and Hutchison 1992), but seasonal changes were seen (Pashkova 1985; Ushakov and Pashkova 1986a,b).

5.5.2
Body Temperature Regulation at High Temperatures

The skin temperature of frogs is generally lower than their core body temperature. However, in small frogs there is hardly any difference (Warburg 1967). The surface tempertature of many frogs is often low due to evaporative cooling. *Chiromantis xerampelina* maintains a body temperature of 2–4 °C below the ambient temperature of 39–43 °C. Thermoregulatory evaporation is controlled by the sympathetic nervous system and the release of endogenous neurotransmitters (Kaul and Shoemaker 1989). *Phyllomedusa sauvagei* tolerates body temperatures of 38–40 °C (Shoemaker et al. 1987).

In salamanders, the body temperature can reach 26.7 °C (average 19.2 °C, see Brattstrom 1963). In anurans, it can reach over 42 °C (Brattstrom 1968). Evaporative cooling, however, seems to be more efficient in anurans compared with urodeles. Thus, *Hyla arenicolor* maintained a body temperature between 28 and 34 °C while in the sun when the ambient temperature was considerably higher (Preest et al. 1992). Likewise, *Bufo marinus* decreased its body temperature by 5.7 °C while losing 42% of its water through evaporation (Malvin and Wood 1991). Both *Litoria caerulea* and *L. chloris* maintained a low EWL up to 40 °C (air temperature). At higher temperatures EWL increased considerably (Buttemer 1990). On the whole, behavioural thermoregulation prevails among most amphibians, including the xeric-inhabiting ones.

In some arboreal anurans exposed to a dry environment, EWL is low (*Hyla cinerea, Phyllomedusa sauvagei, Hyperolius viridiflavus nitidulus*; McClanahan and Shoemaker 1987; Wygoda and Williams 1991; Kobelt and Linsenmair 1995). In *Hyperolius viridiflavus*, the dermal iridiophores serve as a reflective barrier to the sun's radiation (Kobelt and Linsenmair 1986, 1995), reaching 60% reflection (Kobelt and Linsenmair 1992, 1995).

5.5.3
Critical Thermal Maximum

The critical thermal maximum (CTM) is the upper limit of temperature at which the activity of an animal becomes disorganized (Cowles and Bogert 1944). By definition, this state is reversible, and frogs can resume their normal behaviour when temperatures drop. Above this temperature lies the lethal maximum. The CTM of a number of amphibian species, including some xeric-inhabiting ones, have been recorded (Table 5.10).

The CTM depends on the temperature of acclimation (and the habitat), on body mass, and on the stage of development (Delson and Whitford 1973a). It differs in tadpoles compared with post-metamorphic juveniles and adult stages. The CTM of *Bufo marinus* tadpoles is 42 °C whereas that of adults is either similar (Stuart 1951) or 40 °C (Krakauer 1970). Floyd (1983) found that the CTM decreased from a peak of 42–45 to 38 °C at the metamorphic climax. Similarly, *Bufo woodhouseii fowleri* tadpoles had a CTM of 42.5 °C, in juveniles the CTM was 37 °C, and in adults it was 41.1 °C (Sherman 1980a). In *B. marinus* acclimated at 27 °C the CTM was 45 °C (Floyd 1985). In *Ambystoma tigrinum* larvae and neotenes, it was 38–38.5 °C, in post-metamorphic juveniles it was 37–37.5 °C, and in adults it was 36–37 °C (Delson and Whitford 1973a). The CTM of *A. texanum, A. maculatum* and *A. opacum* larvae ranged between 28.1 and 37.9 °C (Keen and Schroeder 1975).

The CTM of *Litoria rubella* was found to vary according to the acclimation temperature and, to some extent, the locality (Warburg 1967). Thus, at higher acclimation temperatures (30 °C) the CTM was higher (40.2 °C) in *L. rubella* from Western Australia, compared with frogs from the Northern Territory. A similar situation was noted in *Ambystoma jeffersonium* (Claussen 1977). Both *Pelobates syriacus* and *Bufo viridis* had a CTM that was 1–1.5 °C higher when acclimated at higher temperatures (Warburg 1971a). The same was true for other amphibian species where the CTM was studied at various acclimation temperatures (see Brattstrom and Lawrence 1962; Brattstrom 1963, 1968,

1970a). Seasonal changes in heat resistance were noted in *Bufo viridis* (Pashkova 1985). In *Liforia caerulea*, the CTM showed a daily variation (Johnson 1971b). *Hyperolius viridiflavus taeniatus* had a CTM of 43 °C when acclimated at 40 °C (Geise and Linsenmair 1988). Altitude also affects the CTM of amphibians (Brattstrom 1968, 1970a). However, these effects of acclimation on CTM are not reflected metabolically, at least as determined in tadpoles of *Limnodynastes peroni* (Marshall and Grigg 1980).

5.5.4
Development Rate and Survival Time at High Temperature

This criterion is of great significance in assessing degrees of adaptation to xeric conditions. Not only is the upper limit of temperature that an amphibian can survive of great importance, but so is the survival time at high temperatures.

The developmental period in amphibians is, to a large extent, dependent upon the temperature (Ushakov and Pashkova 1986a,b). All three species of *Scaphiopus* found in xeric habitats of the southwestern USA (*S. couchi, S. hammondi, S. bombifrons*) complete their development much faster than do the bufonids (*Bufo punctatus, B. cognatus, B. debilis*) which breed in the same habitat (Zweifel 1968). *Bufo valliceps* tolerated up to 35 °C while continuing normal development (Volpe 1957). Brown (1967) found that the later embryonic stages of *S. hammondi* tolerated higher temperatures (39–40 °C) than did the earlier ones. They may complete their metamorphosis even if the temperature ranges between 37 and 39 °C at the gastrulation stage (Brown 1967). The tadpole stages of *S. hammondi* and *S. couchi* wee most heat-resistant (Brown 1969). Among bufonids (*Bufo valliceps, B. luetkeni*), 100% reached the larval stage even at 38 °C (Ballinger and McKinney 1966).

Frogs appear to survive better in dry air when temperatures are comparatively high (Table 3 in Warburg 1965). The leptodactylids *Neobatrachus centralis* and *N. pictus* can survive exposure to dry air at 37.5 °C for more that 8 h (Warburg 1965). The hylid *Litoria rubella* can withstand dry air even at 40 °C for more than 8 h (Warburg 1965, 1967, 1972). Such high temperatures occur in their natural habitat in Central Australia, but normally not for such an extended period. Other frog species from xeric habitats can survive 37.5 °C for the shorter period of 6 h (*Limnodynastes dorsalis*, see Table 3 in Warburg 1965). On the other hand, *Bufo cognatus* survived 40 °C for only 40 min (Schmid 1965a), and *B. valliceps* survived at 38 °C (Ballinger and McKinney 1966).

The chief mechanism of xeric-inhabiting amphibians for surviving high temperatures appears to be their ability to adjust their CTM through acclimation, and thus prolong their survival at high temperatures (Warburg 1967, 1972).

In conclusion, it appears that the main thermal adaptation of amphibians inhabiting xeric environments lies in their longer survival at high temperatures and their thermal preference and CTM, which are higher compared with those of mesic amphibians.

CHAPTER 6

Ecological Adaptations

Any attempt to analyse the ecological adaptations of amphibians inhabiting xeric environments in a comprehensive manner presents a formidable task, especially since amphibians live in two worlds and the factors affecting their lives differ greatly in the aquatic environment compared with the terrestrial-xeric one. (Some of these aspects concerning structure and function have been discussed previously in Chapters 3 and 5.) Foremost among the considerations involved is to decide what aspects need to be reviewed here. How important for its later life as a terrestrial adult are the ecological conditions facing a tadpole or larva in the aquatic environment? Some of these matters were discussed by Wilbur (1980). It appears that most of the experimental work has been carried out on the larval stages, perhaps because they are more accessible than adults and easy to maintain. As it is difficult to decide on the best approach, I shall consider here the ecological adaptations in general of amphibians from xeric habitats to both environments in which they live.

6.1
The Aquatic Environment

Aquatic environments may be of two kinds: a permanent water body (spring, stream, pond or lake) or an ephemeral pool which holds water only for a comparatively short period. The faunas in such water bodies vary because their limnological conditions (minerals, ions, pH, organic matter, oxygen, etc.) are so different. Thus, an amphibian egg in permanent water will probably be exposed to more stable conditions than an egg in a newly formed pond. Consequently, the percentage hatching will probably differ in these two habitats. I am not aware of any study that has attempted to clarify this aspect of differential development in the two types of aquatic habitat. Furthermore, larval life will also differ since the amounts of food as well as the abiotic factors differ in the two habitats (Warburg et al. 1979). We know that salamander (*S. s. infaimmaculata*) larvae hatch in clear water ponds containing very few ions and little organic matter (Warburg et al. 1979). We also know that their growth rate in such ponds differs from their growth rate in either permanent springs or an organically rich pond (Warburg et al. 1979; Degani 1986). Not only is the larval growth rate affected by limnological conditions, but also the time at which larvae metamorphose and consequently the size of the juveniles that emerge onto land. We shall therefore briefly discuss here these matters of larval growth rates and size at metamorphosis under various limnological conditions.

Since the aquatic environment is comparatively homogenous, the life of the larval amphibian is not greatly affected even by drastic variations in the xeric surroundings. Once on land, however, ecological conditions change drastically for all amphibians, especially those inhabiting xeric environments. This is reflected in their activity patterns when exposed to dry and hot air, and in their need to choose suitable shelter for times when they are inactive. The breeding pattern of such species provides the key to their success. These patterns involve the following aspects: the onset of breeding, adult sex ratio, course of reproduction, whether a female is capable of breeding every year as well as how many times per year and how many times in her lifetime, male and female life expectancy, and whether the female is capable of storing sperm. Some of these questions will be briefly discussed here.

6.1.1
Breeding Modes

In amphibians, a variety of reproductive modes are recognized (Goin 1960; Salthe and Mecham 1974; Duellman and Trueb 1986). These range from eggs laid in water with larvae developing in the water, to eggs laid on land with the larvae developing on land (Van Dijk 1971; Tyler 1985). Intermediate modes are illustrated by those anurans that lay their eggs in foam nests on trees over water, with the larvae hatching and falling into the water. Eight reproductive modes can be recognized among Australian frogs: hylids lay their eggs in water, whereas some leptodactylids and myobatrachids lay eggs in foam nests or on land (Roberts 1981; Tyler 1985).

Most anuran species inhabiting xeric habitats lay their eggs in water, and the larvae develop in water (e.g. *Litoria* spp., *Cyclorana* spp., *Neobatrachus* spp., *Heleioporus* spp.). Some lay their eggs in foam nests (*Limnodynastes* spp.), others on land, where the larvae hatch and are washed into pools by the rains (*Pseudophryne* spp.). Finally, some species are known to have a partial or completely terrestrial mode of breeding (e.g. *Arthroleptis, Anhdrophryne, Breviceps, Myobatrachus*; see Van Dijk 1971; Tyler 1985).

Among the ambystomatid urodeles of xeric habitats, egg-laying on both land and in water is known. Other urodeles (*Salamandra salamandra infraimmaculata*) are ovoviviparous (Warburg et al. 1978/1979). Viviparous species give birth to offspring directly. Since these species never form a placenta, they are not considered to be truly viviparous (Wake 1993). Nevertheless, if the larvae hatch from the egg membrane before their birth, such species may be considered to be viviparous (Blackburn 1994). In *S.s. infraimmaculata*, the larvae hatch when the eggs are deposited; rarely are the larvae born alive (Greven 1976; Warburg et al. 1978/1979). This mode of reproduction is characterized by internal fertilization followed by partial development of the ova within the reproductive tract, so that eggs containing fully developed embryos are deposited.

The best strategy for successful breeding in xeric environments appears to be the following: increased longevity of the adult and its delayed maturity coupled with annual breeding (Maiorana 1976). This implies the possibility of mating every year or, alternatively, an arrangement for sperm storage in spermathecae. According to Bragg (1961), the breeding pattern of xeric amphibians is characterized by the following: (1) males precede females when they migrate to the ponds; (2) there are more males than females; (3) breeding starts when water is available. This is usually the situation as regards both anurans and urodeles in xeric environments.

6.1.2
Reproductive Season, Onset and Duration

Two main patterns of reproduction can be recognized among amphibians residing in xeric habitats: first, non-seasonal reproduction in which the animals are capable of breeding at different seasons, provided that an appropriate external stimulus acts as a trigger. The main stimulus is rainfall, but a change in temperature can trigger the onset of reproduction as well. Thus, in *Ambystoma macrodactylum*, rainfall provides the primary stimulus for breeding (Anderson 1967). Likewise, in *S. s. infraimmaculata* the breeding season is triggered by the first heavy rains in November (Warburg 1986a,b, 1992b). In *Ambystoma talpoideum*, cumulative rainfall influences the number of individuals breeding (Semlitsch 1985b). *Batrachoseps attenuatus* as well as two xeric species, *B. major* and *B. relictus*, lay their eggs after the first rainfall of the season (Maiorana 1976).

The other reproductive pattern is strictly seasonal in that it is inflexible. The amphibians with that pattern breed in a certain season irrespective of external conditions. Consequently, the reproductive season is independent of either rainfall or temperature. This type of pattern is seen in *Ambystoma texanum* (Petranka 1984a,b) and in *Scaphiopus intermontanus* (Linsdale 1938 in Hovingh et al. 1985; Sullivan 1985). In some *Scaphiopus* spp., there is no definite breeding season since they are capable of utilizing permanent water (Bragg 1961). In the same way, *Ambystoma tigrinum mavotorium* has no seasonal breeding, unlike all other species of *Ambystoma* (Bragg 1961).

Short-duration breeding periods are known in *Scaphiopus couchi, S. multiplicatus, S. bombifrons, Bufo cognatus, B. alvarius, B. debilis* and *B. reticulatus* (Sullivan 1989). *B. gutturalis* spawns twice during the same season (Balinsky 1985).

The egg-laying period of *Bufo regularis* is during the rainy season: November to January (with 750 mm rainfall and 47 °C ambient surface temperature; Chapman and Chapman 1957). The leptodactylid *Syrrhophus marnocki* breeds throughout most of the year: from February to December (Jameson 1955). In West Africa, *Ptychadena maccathyensis* breeds during March–October, whereas *Ptychadena oxyrhyncus* breeds between December and May (Barbault and Trefaut-Rodrigues 1978). In contrast, *Arthroleptis poecilonotus* has two breeding periods: March to June and August to November (Barbault 1984b).

In Europe, *Salamandra salamandra* breeds from March to September but mostly in July, while larvae are found mainly between March and May (Klewen 1985). On the other hand, in the xeric *S. s. infraimmaculata*, only a single breeding period is known: from November to January (Warburg 1986a, 1994). *Triturus vittatus* males enter water early in the winter, leaving the ponds during February–March, whereas the females leave between February and May (Degani and Mendelssohn 1983).

Breeding late in the season does not necessarily produce larger metamorphs in *Ambystoma talpoideum* (see Semlitsch 1987a). On the other hand, larvae born early in the breeding season have an advantage over those born later. The growth of latecomers may be inhibited by the excrement of older larvae (Rose 1960). In the salamander *S. s. infraimmaculata*, early broods prey upon later broods (Warburg 1992b). In the Amazonian frog *Phyllomedusa tauqteum*, the growth rate of tadpoles born late in the season was adversely affected by the presence of tadpoles belonging to another species (*Osteocephalus taurus*) that had hatched earlier in the same season (Gascon 1992). Likewise, in *Bufo woodhousei* the growth of hatchlings raised alongside larger tadpoles was found to be inhibited, and they survived less well than did the larger tadpoles (Woodward 1987a).

6.1.3
Breeding Migration

The adults of many amphibian species inhabiting xeric environments find a shelter suitable for aestivation on slopes of canyons or in the soil under bushes, sometimes even exposed among the foliage of trees. In many situations rain pools form in depressions or creeks, and amphibians may have to travel considerable distances before arriving there. During this movement, a considerable amount of energy is spent (reviewed by Seale 1987; Bennett 1978). An increase of 8.3 times normal in cardiovascular variables was measured during activity (Withers et al. 1988). Migration to localised breeding sites has been studied largely in amphibians inhabiting mesic environments (Twitty 1959), but hardly in xeric species. There is some evidence, however, that salamanders (*Ambystoma maculatum*) utilize the same track for their migration each year (Shoop 1965). Some aspects of migration will now be discussed briefly.

As with many other species, rainfall controls the breeding migration of *Ambystoma macrodactylum* (Anderson 1967). In *A. talpoideum*, migration reaches a peak during the coldest, not necessarily the wettest months, although it need not respond to temperature (Semlitsch 1985a). *A. tigrinum* males migrate during October–November, and females during February (Semlitsch 1983b). Likewise, in *A. macrodactylum*, 78% of all migration takes place during periods of rainfall, with males appearing first at the ponds (Beneski et al. 1986). A similar phenomenon has been described in *S. s. infraimmaculata*. In this species, male appear first at the ponds during October–November, while most of the females appear during December (Warburg and Degani 1979; Warburg 1986a,b). Neither cold nor cumulative rainfall provides a trigger for breeding migration, but only the first heavy rains after a long, hot, dry season.

6.1.4
Batch or Brood Size

Very little information is available regarding the effects of environmental factors such as rainfall, availability of water, and temperature on anuran egg clutches or urodele brood sizes. These ambient factors are more likely to have an effect on the number of batches or broods than on the number of anuran eggs in a clutch or urodele larvae in a brood.

Among amphibians that inhabit xeric habitats, there is a high variability in either clutch or brood size, ranging from a few eggs to several thousands (Table 6.1). In general, urodeles produce smaller clutches than do anurans. There are, however, exceptions to this in both groups. Thus, *S. s. infraimmaculata* produces on average about 100 larvae per brood (Warburg 1992b, 1994), and *Ambystoma talpoideum* lays 57–504 eggs (Raymond and Hardy 1990), while *Myobatrachus gouldii* deposits 23 eggs in an underground cavity (Roberts 1981). There is also variability in the number of either batches or broods per year, ranging from one annual brood to three or more clutches in *Pachymedusa danicolor* (Iela et al. 1986).

It is generally accepted that a positive relationship exists between brood/batch size and the female's weight (Salthe 1969; Salthe and Duellman 1973; Townsend and Stewart 1994). In *Hyla*, the large clutch size is relative to the female's body size (Crump 1982). The same is true of *Ranidella signifera* (Williamson and Bull 1995) and *Scaphiopus* spp. (Woodward 1987). In *S. s. infraimmaculata* such a relationship has been recently described by Sharon (1995). The relationship between egg size and brood size is less clear (Table 6.2). It seems that even in Amphibia with small clutches, the egg diameter remains rather constant. The same is true in species that lay large batches of eggs.

6.1.5
Development and Metamorphosis

The metamorphic cycle of most anurans and urodeles takes place during the period when the tadpoles or larvae are aquatic. After metamorphic climax, when metamorphosis has been completed, the animals become terrestrial and do not again return to water except to breed. The time spent in water by a terrestrial amphibian has been estimated to be less than 2% of its total life span (Warburg 1988, 1994; Warburg and Rosenberg 1990). Nevertheless, the aquatic

Table 6.1. Clutch size in amphibians from xeric habitats

Family/species	Size	Reference
Anura		
Arthroleptis poecilonotus	20–250 eggs	Barbault and Trefaut-Rodrigues (1979)
Bufo alvarius	8000 eggs	Low (1976)
B. boreas	12 000 eggs per spawn	Samollow (1980)
B. punctatus	3000–5000 eggs	Low (1976)
Cyclorana platycephalus	100–200 eggs per spawn in mucus	van Beurden (1982)
C. australis	7000 eggs	Tyler et al. (1983)
C. longipes	1000–1615	Tyler et al. (1983)
Heleioporus inornatus	Eggs in frothy mass in burrows	Main (1965)
Litoria caerulea	200–2000 eggs in clumps	Tyler and Davies (1986)
L. rubella	715 eggs	Tyler et al. (1983)
Scaphiopus bombifrons	Egg masses 6–110 each 1600	Bragg and Smith (1943) Woodward (1987b)
S. couchi	3310 eggs	Woodward (1987)
S. multiplicatus	1070 eggs	Woodward (1987b)
Phrynomerus bifasciatus	1500 eggs	Rose (1962)
Chiromantis xerampelina	150 eggs	Rose (1962)
Phryobatrachus natalensis	400 eggs	Low (1976)
P. calcaratus	290 eggs	Barbault and Pilorge (1980)
Pyxicephalus delalandii	2500 eggs	Rose (1962)
P. adspersus	3000 eggs	Low (1976)
Ptychadena oxyrhynchus	3476 eggs	Barbault and Trefaut-Rodrigues (1978)
P. maccarthyensis	1330 eggs	Barbault and Trefaut-Rodrigues (1978)
Urodela		
Ambystoma ordinarium	109 eggs deposited singly	Anderson and Worthington (1971)
A. tigrinum	Clutch size related to female's size	Crump (1982); Kaplan and Salthe (1979)
A. texanum	Produces 122 egg masses	Petranka (1984a)
Triturus vittatus	16–68 eggs	Degani and Mendelssohn (1983)

period is critical for an amphibian's post-metamorphic survival, since metamorphosis must coincide with conditions on land that will enable the juveniles to survive in the harsh environments.

Several aspects of amphibian development are of importance in this context, for example, the duration of the aquatic stage and the time spent in water. Supplies of food in the water are likely to influence the feeding patterns of the larval stages, whether they will subsequently become omnivorous, carnivorous or even cannibalistic. Consequently, the duration of time spent as an aquatic larva may have an influence on its survival as an adult and may also affect its size at metamorphosis and, as a result, the size of the juvenile metamorph after metamorphosis. All these aspects will now be discussed briefly.

Table 6.2. Egg diameter of amphibians from xeric habitats

Family/species	Diameter (mm)	Reference
Bufonidae		
Bufo boreas	1.5–1.7	Low (1976)
B. alvarius	1.4	Low (1976)
B. cognatus	1.2	Low (1976)
B. punctatus	1.0–1.3	Low (1976)
B. compactilis	1.4	Low (1976)
B. woodhousei	1.0–1.5	Low (1976)
B. rangeri	1.3	Stewart (1967)
B. carens	1.6	Stewart (1967)
B. carens	1.5–1.9	Balinsky (1957)
B. regularis	1.0	Stewart (1967)
B. regularis	1.4–1.5	Balinsky (1957)
Hylidae *Litoria caerulea*	1.5	Harrison (1922)
Scaphiopidae		
Scaphiopus couchi	1.4–1.6	Low (1976)
S. hammondi	1.0–1.6	Low (1976)
Leptodactylidae		
Nebotrachus kunapalaris	1.6	Mahony and Roberts (1986)
Limnodynastes dorsalis	1.7	Martin (1967)
Heleioporus albopunctatus	2.8	Martin (1967)
H. eyrei	3.3	Martin (1967)
Microhylidae		
Phrynomerus bifasciatus	1.3–1.5	Stewart (1967)
Ranidae		
Breviceps adspersus	4.5	Wager (1986)
Pyxicephalus delalandii	1.5	Wager (1986)
P. delalandii	1.1–1.2	Balinsky (1957)
P. adspersus	2.0	Wager (1986)
P. adspersus	1.6–1.8	Balinsky (1957)
P. natalensis	1.2	Wager (1986)
P. natalensis	0.7–1.2	Balinsky (1957)
Ambystomatidae		
Ambystoma ordinarium	2–8	Anderson and Worthington (1971)

6.1.6
Development Time

The time of development from egg or larva to juvenile varies greatly, even within the same species (Table 6.3). It depends to a certain extent on climatic factors such as rainfall or temperature. Thus, the short duration of ephemeral ponds has been a factor in the evolution of the accelerated growth of *Scaphiopus couchi* tadpoles (Newman 1988b, 1989) and affects the development of *Hyla pseudopuna* tadpoles in drying habitats (Crump 1989). Likewise, the variability in development time in *Heleioporus eyrei* depends on the availability of water.

Temperature also affects the duration of the larval period. In *Ambystoma tigrinum*, the larvae grow faster at increased temperature (Bizer 1978), as do *A. texanum* larvae (Petranka 1984a,b). Likewise, *Scaphiopus couchi* tadpoles develop faster at 30 °C than at lower temperatures (Justus et al. 1977).

Table 6.3. Development time till metamorphosis of tadpoles inhabiting arid habitats

Family/species	Time	Reference
Bufonidae		
Bufo regularis	70 days	Balinsky (1969)
B. carens	41 days	Balinsky (1969)
	29 days till metamorphosis	Stewart (1967)
B. rangeri	35–42 days	Stewart (1967)
B. vertebralis hoeschi	3 weeks development	Channing (1976)
B. cognatus	30–45 days	Stebbins (1954)
	18–45 days to metamorphosis	Krupa (1994)
	28 days minimal duration of larval life	Blair (1975)
B. punctantus	27 days minimal duration of larval life	Blair (1975)
	40–60 days	Stebbins (1954)
B. alvarius	30 days	Mayhew (1968)
	29 days minimal duration of larval life	Blair (1975)
B. arenarum	30 days till metamorphosis, 65 days at low densities	Kehr and Adema (1990)
B. boreas	41–51 days	Karlstrom (1962)
B. canorus	42–52 days	Karlstrom (1962)
Hylidae		
Cyclorana platycephalus	Development time 14 days	Van Beurden (1982)
	30–50 days	Main (1968)
C. cultripes	Larval life 30–50 days	Main (1968)
Litoria rubella	Larval life 40 days	Main (1968)
	Development lasts 14 days	Tyler and Davies (1986)
	Larval period 7 days	Martin (1967)
L. caerulea	2 months until metamorphosis	Harrison (1922)
	38 days	Tyler and Davies (1986)
Leptodactylidae		
Heleioporus eyrei	95–131 days until metamorphosis	Lee (1967)
	Larval life 120–150 days	Main (1968)
H. albopunctatus	128 days until metamorphosis	Lee (1967)
H. psammophilus	141 days until metamorphosis	Lee (1967)
Notaden nichollsi	Larval life 16–30 days	Main (1968)
Neobatrachus sutor	40 days	Main (1968)
N. wilsmorei	40 days	Main (1968)
Pseudophryne occidentalis	30–40 days	Main (1968)
Scaphiopidae		
Scaphiopus couchi	Tadpoles develop in 7 days	Trowbridge (1941a,b)
	Development time 8–16 days	Newman (1989)
	Larval period within 19 days of metamorphosis	Mayhew (1965)
	Tadpoles in 2–6 weeks	MacMahon (1985)
	12–13 weeks	Bragg (1965)
	18–28 days	Low (1976)
S. hammondi	27–36 days to metamorphosis	Blair (1975)
	51 days	Low (1976)
S. bombifrons	Development until metamorphosis 32 days	Trowbridge (1941a,b)
	2 months	MacMahon (1985)
	36–40 days	Low (1976)
S. holbrooki	Development within 21 days	Richmond (1947)

Table 6.3 (*Contd.*)

Family/species	Time	Reference
Microhylidae		
Phrynomerus annectens	8 weeks of development	Channing (1976)
P. bifasciatus	30 days	Stewart (1967)
Ranidae		
Tomopterna delalandii	5 weeks until metamorphosis	Jurgens (1979)
Pyxicephalus delalandii	35 days	Stewart (1967)
	36 days	Balinsky (1969)
P. adspersus	28–42 days	Wager (1986)
	35 days	Low (1976)
	47 days	Balinsky (1969)
	49 days	Stewart (1967)
Phrynobatrachus	29 days	Low (1976)
natalensis	31 days to metamorphosis (at 30 °C)	Balinsky (1957)
	49 days	Stewart (1967)
Ambytomatidae		
Ambystoma texanum	9–12 days to metamorphosis	Petranka (1984a,b)
A. tigrinum	49 days	Petranka (1989)
	9–12 days till metamorphosis	Petranka (1984a,b)

In *A. texanum*, larval growth is faster in ponds than in streams, and consequently, the larval period was shorter in ponds (Petranka 1984a,b). Pond size also influences the duration of the larval period, and the survival and metamorphosis of the tadpole (Pearman 1995). It also has an effect on tadpole competition (Pearman 1991). Tadpoles of *Hyperolius viridiflavus* reared at low densities metamorphosed earlier (Schmuck et al. 1994).

Development time may also depend upon biotic factors such as ovum size, availability of food and density of the tadpoles (Kaplan 1980a,b; Wilbur 1980; Tejedo and Requer 1994). Thus, low food levels caused a decreased average level of growth rate in *Hyla gratiosa* tadpoles, resulting in a prolonged larval period (Travis 1984). Likewise, in *Hyla cinerea* food shortage induces a longer developmental period (Leips and Travis 1994). Food levels regulate the size at metamorphosis in *Rana tigrina* (Hota and Dash 1981). Conversely, an ample food supply enhances the growth of *Salamandra* larvae (Degani 1993). Tadpoles of *Hyperolius* spp. reared at low densities metamorphosed sooner and at a larger size than others reared at high densities (Schmuck et al. 1994). In *Scaphiopus couchi* high density causes early metamorphosis although at a smaller size (Newman 1994).

Finally, the pH of the water also affects the growth of amphibian larvae. Thus, at the beginning of the season when pond water has a lower pH (= 6.5), salamander larvae (*S. s. infraimmaculata*) grow more slowly than later in the season when the pH of the pond is higher (= 8.7; Warburg et al. 1979).

The development time may also vary according to the locality. Thus, the development time in *Scaphiopus couchi* ranged from 7 days (Trowbridge

1981a,b), 8–16 days (Newman 1989), 12–14 days (Bragg 1961, 1965) to 19 days (Mayhew 1965) in different places. In other anuran species, the duration of development also shows a great variability: *Bufo regularis* 75 days, *B. carens* 37–52 days, *Phrynobatrachus natalensis* 27–40 days, *Kassina senegalensis* 52–64 days, *Cacosternum boettgeri* 37 days, *Pyxicephalus delalandei* 25–35 days, and *Pyxicephalus adspersus* 32 days. The latter was successful in breeding only once in 5 years, when the pond remained there for a long enough period to enable the tadpoles to complete their metamorphosis (Balinsky 1969).

In the urodele *Ambystoma opacum*, the development time is about 42 days (Petranka and Petranka 1980), and in *A. maculatus* it is between 57 and 144 days (Wilbur and Collins 1973). In *S. s. infraimmaculata* the average development lasts 56 days (Warburg et al. 1978/1979, 1979).

Further data on development time are given in Table 6.3. It is impossible to formulate any hypothesis or to make any prediction as to the probable development time of any particular species. Perhaps the only valid generalization to be made is that some desert species have shorter development times than species from mesic environments. On the other hand, it would not be correct to state that species adapted to xeric environments have, in general, an aquatic larval life of short duration since some of them have unusually long aquatic larval periods.

Some of the characteristics of the aquatic stage of amphibians that inhabit xeric habitats concern feeding patterns (Bragg 1961). Thus, *Scaphiopus* tadpoles aggregate while feeding. Some of them become predaceous or even cannibalistic. The latter phenomenon has been observed both in anurans and urodeles (Warburg 1992b).

6.1.7
Larval Competition for Food Resources, Carnivory, Predation and Cannibalism

Anuran tadpoles are usually omnivorous, feeding largely on vegetable material, but they are occasionally also carnivorous (see review by Crump 1992). On the other hand, urodele larvae are invariably carnivorous and occasionally become cannibalistic. Although amphibian larvae do not usually compete for resources, food may become scarce as time advances because developing tadpoles or larvae require larger quantities of food. Competition may therefore arise. A shortage of food may adversely affect the rate of differentiation, as in tadpoles of *Bufo woodhousei* (Alford and Harris 1988); it can also reduce the survival of *Ambystoma texanum* larvae and lengthen their development time to metamorphosis (Petranka 1984a,b), as well as their size at metamorphosis, which is adversely affected by larval density (Petranka and Sih 1986).

The proportion of metamorphosing *Rana tigrina* tadpoles was higher at low densities (Dash and Hota 1980). *Scaphiopus couchi* tadpoles raised at low density developed quickly, and metamorphosis was not observed at high density (Newman 1987). Density affects growth, time to metamorphosis and mortality in *Bufo americanus* (Brockelman 1969).

Competition between *Bufo* and *Hyla* tadpoles for food resources was found to be stronger when *Hyla* tadpoles hatched in the ponds 7 days after *Bufo* tadpoles (Lawler and Morin 1993). Consequently, the presence of *Bufo* tadpoles significantly depressed the weight of the metamorphosing young *Hyla* juveniles (Morin et al. 1988). There is obviously a great advantage in spawning early in the season (Wilbur and Alford 1985).

At low densities, *Hyla arborea* tadpoles became larger than when they were more crowded (Grosse and Bauch 1988). Likewise, when larvae of *A. opacum* were raised at low densities, they metamorphosed to become larger adults (Scott 1994). High density adversely affected their growth (Stenhouse et al. 1983).

Among other things, high larval density caused increased competition for food in *Scaphiopus multiplicatus* tadpoles and resulted in carnivory (Pfennig 1990). Cannibalism among *S. s. infraimmaculata* larvae apparently increases with growth and with the availability of food (Degani et al. 1980). On the other hand, in the presence of a large supply of invertebrate food, no cannibalism was observed among *S. s. infraimmaculata* larvae, whereas a shortage of food induced cannibalism (Degani 1993; see also Fig. 6.1). The intensity of competition for food among *Scaphiopus* spp. tadpoles is also related to pond size (Woodward 1982, 1987a). Woodward (1983) found that more predation took place in permanent than in temporary ponds. The high density of *Ambystoma tigrinum* larvae (when food is limited) also affects the frequency of cannibalism (Crump 1983; Collins and Cheek 1983; Lannoo and Bachmann 1984; see reviews by Kuzmin 1991; Elgar and Crespi 1992), like in *Pseudacris* frogs (Loeb et al. 1994). On the other hand, the frequency of cannibalism is not always correlated significantly with larval density (Loeb et al. 1994).

Three different kinds of morphs have been observed among *Ambystoma tigrinum* larvae: neotene, small morphs and broad-headed cannibals (Rose and Armentrout 1976). The large, broad-headed larvae of *A. annulatum* are also cannibalistic (Nyman et al. 1993). These broad-headed morphs have been found among both sexes (Collins and Holomuzki 1984). In *Ambystoma macrodactylum*, a different diet enhanced morphological variations (Walls et al. 1993). Apparently, differentiation into the broad-headed morphs begins early in ontogeny (Pierce et al. 1983). Furthermore, some difference in genetic frequency was found between cannibals and non-cannibals (Pierce et al. 1981).

In ponds inhabited by both *Scaphiopus couchi* and *S. bombifrons*, tadpoles belonging to the latter species prey upon the tadpoles of the former species without showing any cannibalistic tendency towards their own siblings (Bragg 1956, 1962). Tadpoles of *S. multiplicatus* can change from omnivorous to carnivorous feeding and then back again to omnivorous, depending on their supplies of food (Pfennig 1992a,b). They can also change their feeding pattern when exposed exogenously to thyroxin. They then assume the morphology of carnivorous larvae (Pfennig 1992a,b). Cannibalistic *Ambystoma tigrinum* larvae metamorphose sooner than omnivorous ones (Lannoo et al. 1989). Like-

Fig. 6.1. *Salamandra* tadpoles living under a stone on wet mud after the pond had dried up (*top* and *middle left*). The surviving larvae will become cannibals (*middle right*) and feed on later cohorts. A large adult specimen is shown (*bottom*)

wise, *S. multiplicatus* cannibals develop rapidly and metamorphose sooner than omnivorous tadpoles (Pfennig et al. 1991). There is a report that *Hyla pseudopuna* tadpoles, when fed on conspecifics, grew faster than when fed on the tadpoles of other species (Crump 1990).

6.1.8
Survival of Larvae

The survival of larvae is to a large extent dependent upon the persistence of the pond in which they are living. If the water evaporates and is not replenished by rain in time, the larvae will perish (Fig. 6.1). This could result in the loss of an entire cohort and sometimes means no successful breeding during a particular year. Larvae may be saved if the break between rainfall episodes is not too long. This was observed in *S. s infraimmaculata* larvae, which can survive under stones on wet soil for about 1 week (Warburg 1986a,b). It was calculated that in most years only 3.3% *Ambystoma tigrinum* larvae survive (Anderson et al. 1971).

Survival of larvae depends also on various other factors. The percentage survival of *Ambystoma talpoideum* larvae hatching from larger eggs is greater than that of larvae from small eggs (Semlitsch and Gibbons 1990). The reason for this could be the fact that the former grew faster and metamorphosed earlier (Williamson and Bull 1989, 1992). However, in some ambystomatids ovum size has no effect on embryo development time to hatching stage of the tadpole (Kaplan 1980b), although it appears to affect the size of *Ambystoma opacum* larvae (Walls and Altig 1986) and of *Ranidella* spp. tadpoles (Williamson and Bull 1989). The survival of *A. maculata* larvae is not dependent on the number of eggs in the batch, but on their rate of development as related to the persistence of the pond (Shoop 1974).

There are conflicting reports about the relationship between larval density and survival. Thus, low larval density resulted in better survival in *A. opacum* (Scott 1994). On the other hand, the survival of *Bufo americanus* (Wilbur 1977) and *Hyoperolius viridiflavus* tadpoles (Schmuck et al. 1994) was independent of their density.

6.1.9
Factors Affecting Success of Metamorphosis

It is generally assumed that metamorphosis in Amphibia takes place once a certain minimum size has been attained (Wilbur and Collins 1973). Accumulation of energy in the tadpole stage is a requirement that ensures successful metamorphosis (Pandian and Marian 1985).

S. s. infraimmaculata larvae metamorphosed earlier in the season than did those of other salamander species in ephemeral ponds (8–10 weeks) or than larvae of the same species that inhabit permanent pools (5 months; Warburg et al. 1979). In *Ambystoma tigrinum*, both the time to metamorphosis and body size are partially correlated with the date on which the pond dries up (Semlitsch et al. 1988).

The size of *Hyla pseudopuna* metamorphs is reduced if the pond evaporates (Crump 1989). This has also been described in *Scaphiopus couchi* tadpoles (Newman 1988a,b, 1989), as well as in other anuran species (Wilbur et al. 1983) which metamorphosed at a smaller size when the ponds dried out. The amount of rainfall has been shown to control the size of *Ambystoma macrodactylum* at metamorphosis (Anderson 1967).

The number of *Bufo americanum* reaching metamorphosis is density-dependent (Wilbur 1977), but their size at metamorphosis is largely influenced by their growth rate (Travis 1984). Only 3% of *Ambystoma texanum* larvae survived successfully till metamorphosis, but their survival was not necessarily related to any density factor (Petranka and Sih 1986). Low food levels caused *Hyla gratiosa* tadpoles to metamorphose at a smaller size (Travis 1984). Ponds drying up also affected the number of *A. talpoideum* tadpoles metamorphosing (Semlitsch 1987a,b; Semlitsch and Wilbur 1988).

There are conflicting reports as to the effect of temperature on size at metamorphosis. *A. tigrinum* larvae metamorphose at a smaller size when growing at high temperatures (Bizer 1978; Sexton and Bizer 1978), whereas *A. texanum* larvae become larger when they metamorphose at a higher temperature (Petranka 1984b). Size appears to have little effect on the dehydration tolerance of *Scaphiopus couchii* tadpoles (Newman and Dunham 1994).

Interspecific differences in growth rate under similar conditions are also found. Thus, *A. opacum* larvae grew faster than *A. maculatum* larvae, and these in turn grew faster than *A. talpoideum* larvae raised under similar conditions (Walls and Altig 1986).

6.1.10
Reproductive Strategies

Although much is known about the length of their annual breeding season, scarcely anything is known about the number of times that a single female can reproduce. The European *Salamandra salamandra* is believed to have a biannual breeding season (Joly et al. 1994). On the other hand, in the xeric *S. s. infraimmaculata* there is only one breeding season each year (Warburg 1992a; Sharon 1995). It is not known, however, whether each salamander female reproduces annually, biannually or perhaps only once every few years. In one long-term study of *S. s. infraimmaculata*, in which individual females were captured year after year for over 20 years, it was found that they visited the breeding ponds on several consecutive years (Warburg 1996; Fig. 6.2).

In *Cyclorana platycephalus* the female does not spawn her entire batch on a single occasion. Some of the eggs are stored over the winter (Van Beurden 1979). This cannot happen in *S. s. infraimmaculata*, however, because all the larvae in the uterus are released at the same time (Warburg et al. 1978/1979). Likewise, almost nothing is known about the percentage of breeding females within a population. Although this is extremely difficult to study, at least some estimate can be obtained regarding the percentage of females in a population which contain post-vitellogenic ova that are likely to ovulate during the next breeding season.

It is known that salamanders are capable of storing spermatophores in spermatheca located in the roof of the cloaca (Sever 1994; Sever et al. 1995). Yet we have evidence that *S. s. infraimmaculata* returns year after year to the same pond to breed. Since there are always males present, they undoubtedly can and do mate on such occasions, and consequently do not need to store sperm in the spermathecae. In some studies, though, sperm was found in the cloaca of every salamander female even before she had reached a pond, and the sperm was probably viable (Joly et al. 1994; Greven and Guex et al. 1994). It is possible that sperm storage in spermathecae ensures that the female reproduces in case mating does not take place. On the other hand, multiple mating can cause sperm mixing. In *Chiromantis xerampelina* multiple mating was shown to take place, and consequently there is sperm competition (Jennions and Passmore 1993).

6.1.11
Habitat Use

Very few studies have dealt with the ways in which an aquatic habitat is used simultaneously by more than one amphibian species. Thus, in an ephemeral pond inhabited by the tadpoles and larvae of six amphibian species, one species was found to be an early breeder, one species a spring breeder, one a summer breeder, while three species bred during both the spring and summer (Degani 1982c). Consequently, the larval stages of at least four species were present concurrently in the pond throughout most of its duration. Blair (1961) describes how seven anuran species shared one particular breeding pond. One species bred in the winter, three in the spring, two in the summer, and one had an extended breeding period from the winter through spring. In a pond inhabited by nine amphibian species, Diaz-Paniagua (1992) found two species breeding early in the season, while one species was a late breeder. The others all bred during the same period.

Amphibian larvae and tadpoles of different species seem to utilize different microhabitats and food (Degani 1986). Thus, in one pond, one species may be

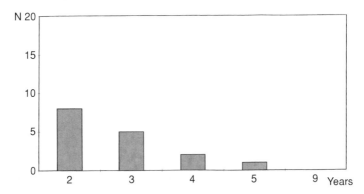

Fig. 6.2. Pond fidelity in *Salamandra* females on Mt. Carmel. Eight female salamanders visited the same pond for 2 consecutive years, five for 3 consecutive years, two for 4 consecutive years and a single female year after year for 5 consecutive years

sedentary and four species bottom dwellers, while one utilizes the entire pond space. A similar partitioning of the aquatic habitat has been described for four anuran species by Alford (1986), as well as in several African anurans (van Dijk 1972). Another way of utilizing the different microhabitats in the same pond has been described in *Ambystoma talpoideum* larvae which migrate vertically (Anderson and Williamson 1974).

In a mixed population of tadpoles belonging to two species of *Ranidella*, segregation was seen in the use of the microhabitat. One species was commonly found on rocks, but in mixed populations it increased its presence on the substrate between the rocks (Odendaal et al. 1982).

6.2
The Terrestrial Amphibian

The ecological conditions under which xeric-adapted adult amphibians live vary from extremely arid during most of the year to mildly xeric during the remainder (Table 6.4). Furthermore, in some arid and semi-arid habitats, the rainy season takes place during the warm summer months. This changes the ecological conditions and permits the survival of amphibians, among other taxa, that would otherwise not be able to endure the long, hot summer. On the other hand, in some of the mildly xeric habitats, especially in the Mediterranean region, most of the year is dry except for the short winter period when it rains. Mild temperatures prevailing during part of the winter enable breeding to take place. However, the amphibians inhabiting these habitats have to adapt to the long, dry summer months (Table 6.5).

When discussing the ecology of adult Amphibia we must take into consideration the fact that they exist in three different forms: (1) as adults that remain in water but retain their larval character – the neotenous type; (2) as adults that remain either completely aquatic or spend most of the time in water; (3) terrestrial species, some of which have become adapted to xeric environments. Among the latter, some return to water only to breed.

The physiological adaptations of the first two types have been discussed previously (Warburg and Rosenberg 1990). Therefore, they will not be discussed here in any detail. Even if some species also inhabit semi-arid or arid regions, they do not possess, as far as is known, any specific adaptations enabling them to inhabit water bodies in the desert. However, too little is known about this interesting point. The third type, the terrestrial amphibian, is the subject of our discussion. Of this type only a few species have so far been studied, and some of those adapted to xeric conditions will be mentioned here.

During the life of terrestrial amphibians we can distinguish between two principal phases, the terrestrial and the aquatic. Adult amphibians returning to water in order to breed really belong to the latter phase. In some forms, the aquatic phase may be prolonged – lasting for several weeks – during which the

Table 6.4. Regions and habitats inhabited by amphibians adapted to xeric conditions

Species	Region/habitat	Reference
African amphibians		
Bufo pentoni	Sahelian savannah	Forge and Barbault (1978)
	Dry savannah	Schiotz (1967)
	Sub-desert	Francillon et al. (1984)
B. regularis	Sahara	Lambert (1984)
B. viridis	Hoggar, Central Sahara	Mayhew (1968); Lambert (1984)
B. garipensis	Karoo, South Africa	Rose (1962)
B. carens	Semi-arid	Balinsky (1969)
B. mauritanicus	Sahara	Lambert (1984)
Hyperolius viridiflavus nitidulus	Dry savannah	Schiotz (1967)
H. nasutus	Dry savannah	Schiotz (1967)
H. pusillus	Dry savannah	Schiotz (1967)
Kassina senegalensis	Dry savannah	Schiotz (1967, 1976)
K. cassinoides	Dry savannah	Schiotz (1967)
Leptopelis bufonides	Dry open savannah	Schiotz (1967)
L. viridis	Savannah	Schiotz (1967)
Afrixalus pygmaeus	Dry savannah	Schiotz (1976)
A. farnasinii	Savannah	Schiotz (1976)
Breviceps verrucosus	Kalahari savannah	Poynton (1964), Poynton and Pritchard (1976)
B. gibbosus	Arid	Rose (1962)
B. mossambicus	Open grassland	Stewart (1967)
B. bifasciatus	Open grassland	Stewart (1967)
B. adspersus	Savannah, Kalahari	Poynton (1964); Poynton and Pritchard (1976)
B. fowleri	Open woodland	Stewart (1967)
Cacostermum bottgeri	Arid	Poynton (1964)
Ptychadena oxyrhychus	Savannah	Barbault and Trefaut-Rodrigues (1978)
P. maccathyensis	Savannah	Barbault and Trefaut-Rodrigues (1978)
P. owleri	Tropical Sahara	Lambert (1984)
Arthroleptella lightfooti	Mountains	Rose (1962)
Phrynobatrachus natalensis	Lowland savannah	Stewart (1967)
Pyxicephalus dellalandii	Open grassland, arid	Poynton (1964); Coe (1974)
P. adspersus	Open grassland, arid	Stewart (1967)
Tomopterna delalandii cryptotis	Semi-arid, sahelian savannah	Schiotz (1967); Loveridge (1976); Channing (1976); Forge and Barbault (1978); Jurgens (1979)
Chiromantis xerampelina	Dry woodlands, arid	Coe (1974)
C. petersii	Semi-arid savannah	Coe (1974)
C. rufescens	Arid	Coe (1974)
Australian amphibians		
Cyclorana platycephalus	Arid, clay soil	Main (1965); Tyler and Davies (1986); Tyler et al. (1981)
C. cultripes	Arid	Glauert (1945); Main (1965); Tyler et al. (1981)
C. cryptotis	Arid, clay soil	Tyler and Davies (1986); Tyler et al. (1982)
C. maini	Arid	Glauert (1945)
C. verrucosus	Arid	Glauert (1945)

Table 6.4 (*Contd.*)

Species	Region/habitat	Reference
C. vagitus	Grassland	Tyler and Davies (1986)
Litoria rubella	Rocks	Main (1965)
L. caerulea	Woodland	Main (1968)
L. peronii	Semi-arid	Cogger (1975, 1992); Tyler et al. (1981)
Heleioporus eyrei	Sandy soil	Main (1965)
H. australiacus	Granite	Main (1965)
Neobatrachus sutor	Desert clay pans	Main (1965, 1968); Main et al. (1959)
N. pelobatoides	Rocky habitats	Bentley et al. (1958)
	Clay pans	Main (1968)
N. wilsmorei	Arid, clay pans	Main (1968); Main et al. (1959); Bentley et al. (1958)
N. centralis	Desert clay pans, mallee	Main (1968); Main et al. (1959)
Notaden nichollsii	Spinifex sand hills	Main (1968)
N. bennetti	Savannah	Cogger (1975)
Limnodynastes spenceri	Arid	Cogger (1975); Tyler et al. (1981)
L. ornatus	Most xeric	Tyler et al. (1981)
Pseudophryne bibroni	Dry sclerophyl forest	Cogger (1975)
P. occidentalis	Arid	Main et al. (1959)
Uperolia rugosa	Arid	Cogger (1975)
Glauertia russelli	Desert	Main et al. (1959)
Myobatrachus gouldii	Sandy soil	Main (1965)
South American amphibians		
Bufo chilensis	Arid acacia steppe	Cei (1980)
Phyllomedusa sauvagei	Arid, xerophyl scrub, Chaco	Cei (1980)
Pleurodema nebulosa	Salt flats	Cei (1980)
P. tucumana	Salt flats	Cei (1980)
P. thaul	Arid acacia	Cei (1980)
Lepidobatrachus llanensis	Arid, Chaco	Cei (1980)
L. asper	Arid, Chaco	Cei (1980)
Leptodactylus bufonius	Arid	Cei (1980)
L. chaquensis	Arid	Cei (1980)
Physalaemus albonotatus	Chaco	Cei (1980)
P. nattereri	Arid	Cei (1980)
Odontophrynus occidentalis	Sandy habitat	Cei (1980)
Alsodes nodosus	Arid, acacia, steppe	Formas (1979)
Ceratophrys pierotti	Very dry regions	Cei (1980)
North American amphibians		
Bufo cognatus	Grass prairie, Chihuahua	MacMahon (1985); Bragg and Smith (1943); Creusere and Whitford (1976)
B. americamus	Savannah	Bragg (1944)
B. microscaphus	Gravel	MacMahon (1985)
B. compactilis	Short grass plains, Chihuahua	Morafka (1977); Bragg and Smith (1943)
B. punctatus	Rocks, prairie grassland, arid	MacMahon (1985); Bragg and Smith (1943); Tevis (1966)
B. woodhousei	Desert grass	Bragg and Smith (1943)

Table 6.4 (*Contd.*)

Species	Region/habitat	Reference
B. alvarius	Desert grassland	MacMahon (1985); Low (1976)
B. boreas	Desert	Miller and Stebbins (1964)
B. speciosus	Chihuahua desert	MacMahon (1985); Morafka (1977)
B. debilis	Rocks, prairie semi-arid	MacMahon (1985)
Hyla cadaverina	Desert	MacMahon (1985); McClanahan et al. (1994)
H. arenicolor	Rocks, arid	MacMahon (1985)
H. eximia	Chihuahua	Morafka (1977)
H. californiae	Rocky canyons	Miller and Stebbins (1964)
Pternohyla fodiens	Arid, mesquite	MacMahon (1985)
Eleuterodactylus august	Chihuahua, mesquite	Morafka (1977)
E. latrans	Semi-arid, arid	Low (1976)
Syrrhopus marnocki	Rock crevice, Chihuahua	Cochran (1961); Morafka (1977)
Scaphiopus couchi	Chihuahua, arid	Creusere and Whitford (1976); Mayhew (1962, 1965); Bragg (1944)
S. hammondi	Xeric, semi-arid region, Chichuhua	Bragg (1944); Creusere and Whitford (1976); MacMahon (1985); Ruibal et al. (1969)
S. intermontanus	Sagebrush flats, xeric habitats	Bragg (1944); MacMahon (1985)
S. bombifrons	Mixed grass prairie, Chihuahua	Bragg (1944); MacMahon (1985); Creusere and Whitford (1976)
S. hurterii	Xeric savannah	Bragg (1944)
Gastrophryne olivacea	Chihuahua	Morafka (1977); MacMahon (1985)
Ambystoma tigrinum	Arid sagebrush plains	MacMahon (1985); Collins (1981)
Batrachoseps campi	Arid	Yanev and Wake (1981)
B. aridus	Arid	Tevis (1966)

Table 6.5. Life habits of amphibians inhabiting xeric habitate

Family/species	Ecological conditions	Reference
Anura		
Bufo arenarum	Burrows	Blair (1960)
B. spimulosus	Volcanic rocks	Cei (1980)
B. cognatus	Volcanic rocks	Blair (1960); McMahon (1985)
B. punctatus	Rock crevices	Blair (1960)
B. woodhousei	Burrows	Blair (1960)
B. carens	Away from water	Balinsky (1969)
B. regularis	Cultivated land, garden near river	Balinsky (1969)
Hemisus marmoratus	Burrows	Rose (1962)
Cyclorana platycephalus	Burrows 30 cm	Main et al. (1959)
C. cultripes	Burrows	Main (1965); Tyler et al. (1981)
C. cryptotis	Fossorial	Tyler et al. (1982)
Hyperolius marmoratus	On branches	Coe (1974)
H. viridiflavus	Dry plants, unshaded places	Geise and Linsenmair (1986, 1988)

Table 6.5 (*Contd.*)

Family/species	Ecological conditions	Reference
Hylambates maculatus	Aestivates between leaves	Rose (1962)
Afrixalus farnasinii	Between leaves	Rose (1962)
Eleutherodactylus latrans	Fossorial	Low (1976)
Lepidobatrachus llanensis	Fossorial	Cei (1980)
L. asper	Fossorial	Cei (1980)
Ceratophrys ornata	Fossorial	Cei (1980)
Telmatobius reverberii	Under trees	Cei (1980)
T. solitarius	Volcanic rocks	Cei (1980)
Pleurodema nebulosa	Fossorial	Cei (1980)
Physalaemus nattereri	Fossorial	Cei (1980)
Odonthophrynus americanus	Fossorial	Cei (1980)
Breviceps verrucosus	Fossorial	Loveridge (1976)
B. adspersus	Burrows 51–60 cm deep	Poynton and Pritchard (1976)
B. adspersus	Burrows 180 cm deep	Rose (1962)
Phrynomerus bifasciatus	Burrows	Rose (1962)
Chiromantis xerampelina	Exposed places	Poynton (1964); Rose (1962)
Leptopelis bocagei	Fossorial	Rose (1962); Coe (1974)
Heleioporus eyrei	Sand	Main (1965)
H. albopunctatus	Burrows 67–115 cm deep	Lee (1967)
H. albopunctatus	Burrows 70–80 cm deep	Bentley et al. (1958)
H. inornatus	Sand	Main (1968)
H. psammophilous	Clay	Main (1965)
H. australiacus	Burrows	Main (1965)
H. barycragus	Fossorial	Lee (1967)
Neobatrachus pelobatoides	Rocks	Bentley et al. (1958)
N. wilsmorei	Fossorial	Main (1965)
Notaden nichollsii	Burrows 122 cm	Main (1965); Mayhew (1968)
N. nichollsii	Burrows 170 cm	Slater and Main (1963)
Limnodynastes dorsalis	Burrows	Main (1968); Tyler et al. (1981)
Myobatrachus gouldii	Fossorial	Main (1965)
Scaphiopus couchi	Burrows 61 cm	Mayhew (1968); Bragg (1944)
S. hammondi	Fossorial, 91 cm	Bragg (1944); Ruibal et al. (1969)
S. intermontanus	Fossorial	Bragg (1944)
S. bombifrons	Fossorial	Bragg (1944, 1945)
Pyxicephalus adspersus	Fossorial	Balinsky (1969)
P. delalandii	Fossorial	Coe (1974); Loveridge (1976)
P. natalensis	Fossorial	Coe (1974); Loveridge (1976)
Tomopterna delalandii	Fossorial	Channing (1976); Loveridge (1976)
Urodela		
Ambystoma talpoideum	Semi-permanent ponds	Semlitsch (1985b)
A. tigrinum	Sagebush plains	MacMahon (1985)
Siren lacertina	Burrows	Etheridge (1990b)
S. intermedia	Burrows	Gehlbach et al. (1973)

animal undergoes both structural and physiological changes. These changes have recently been examined (Warburg et al. 1994b; Warburg and Rosenberg 1990). Some of the ecological adaptations of terrestrial, xeric-adapted forms will now be discussed.

6.2.1
Aquatic Phase of the Adult Amphibian Inhabiting Xeric Habitats

6.2.1.1
The Paedomorphic Amphibian

Paedomorphosis is a phenomenon in which sexually mature adults remain in water in their larval form. This phenomenon has been described in a number of urodeles, particularly in *Ambystoma* spp. (Collins 1979). In *A. rosaceum*, neoteny is dependent on permanent water (Anderson and Webb 1978). A particularly high percentage of facultative paedomorphic salamanders (*A. talpoideum*) maintain low density populations in permanent water bodies (Semlitsch and Wilbur 1989). Food levels did not appear to affect the percentage of paedomorphs (Semlitsch 1987c). Paedomorphic salamanders (*A. dumerilii*) may undergo spontaneous metamorphic changes in the lab (Brandon 1976). The larvae of paedomorphic *A. talpoideum* can become sexually mature at the same age as normal animals, but they are smaller. This difference in size is also maintained in 1-year-old paedomorphs which produce fewer eggs than normal ones do (Semlitsch 1985b). On the other hand, paedomorphic adults of *A. talpoideum* laid their eggs 6 weeks earlier than did normal adults (Scott 1993). Paedomorphosis is believed to be of some advantage to both salamanders and newts (Whiteman 1994; Scott 1993). Among sexually mature larvae the sex ratio differed from that found among individuals that had metamorphosed in the normal way and more females than males were present.

A valid explanation of this phenomenon is still lacking, and much experimental work needs to be carried out in order to determine under what conditions paedomorphosis can be induced.

6.2.2
The Adult Terrestrial Amphibian Inhabiting Xeric Habitats

6.2.2.1
Ecological Problems with Breeding

The Availability of Water. Most amphibians inhabiting xeric habitats need water for breeding purposes. Moreover, the water must be available for a period long enough to enable metamorphosis to be completed. This period varies in different species ranging from 7 days (in *Scaphiopus* sp.) to 8 weeks (in *Pelobates syriacus, Salamandra salamandra*). Some of the amphibians breed in ephemeral ponds, entering them to spawn just after the pond has been formed, whereas others will spawn only when the aquatic plants which the female needs for spawning have started to grow (*Triturus vittatus*). To the first group belong many desert frogs which generally have short metamorphic periods, while to the other group belong some of the urodeles which have a longer larval life. However, in *Ambystoma texanum*, two populations differ in their breeding sites and egg size (Petranka 1984c). *Pyxicephalus adspersus* spawns in very shallow pools which are likely to dry out rapidly. The nest contains eggs which are protected by the male (Balinsky and Balinsky 1954). Several other terrestrial amphibian species are also known to protect their nests (Rabb 1973).

Reproductive Strategies. Due to the unpredictable nature of precipitation in their environment, amphibians that inhabit xeric habitats and are dependent on water during their larval life show flexibility in the duration and frequency of their reproduction. The onset of reproduction is totally dependent on the presence of water. However, how a female amphibian senses the correct date and is ready with post-vitellogenic oocytes is by no means clear. Moreover, in several species, the time available for larval development is somewhat limited. Thus, any delay in the onset of reproduction is critical since the larvae may not have sufficient time in which to develop. This brings us to the second point: the duration of the reproductive period. Here we can imagine two scenarios. First, the female breeds once during her reproductive period. In that case she will miss her opportunity if her reproductive cycle is not synchronized with the onset of precipitation, or if the larval period is so extended that it requires water for longer than any is available. Another option open to the female is to have an additional breeding period during the same season. In that case it is less essential for the female to be absolutely ready for the rainy season. It can wait for the first rains to trigger the onset of the vitellogenic cycle. However, it may thus risk the loss of its cohort, due to evaporation. If this occurs, the question remains, why did amphibian females adopt such an inflexible mode of reproduction? The selective pressure of such a harsh environment should theoretically have eliminated rigid reproductive patterns in favour of more flexible ones. This brings up the last major point: frequency of reproduction. It would appear that reproducing frequently, or at least several times during an animal's lifetime, would be advantageous given the unpredictable weather conditions found in xeric environments.

   Except for a single, long-term study on salamanders (*S. s. infraimmaculata*), in which the breeding pattern of individual salamanders was followed over several years, there are no comparative data. In that study (Warburg 1996), it was found that single salamander females bred year after year for several years (Fig. 6.2).

6.2.2.2
Role of Climate in the Survival of Juveniles and Adults

Drought can be critical for young post-metamorphs if they do not find suitable refuges in time. Even adults can be overcome by harsh, hot and dry spells. Thus, specimens of *Batrachoseps atteniatus* have been found dying in the field, largely due to heat stress and perhaps also from desiccation (Maiorana 1977). A persistent drought of a few years' duration may cause amphibians to remain in their hiding places underground. Several of the amphibians that inhabit xeric environments are long-lived and thus can afford to skip one or more reproductive periods (Tevis 1966). There is evidence that *S. s. infraimmaculata* females on Mt. Carmel return year after year, sometimes breeding unsuccessfully (Warburg 1988, 1992b). Based on the rainfall pattern during the period, an estimation has shown that, on average, about half of that species' reproductive efforts could result in successful metamorphosis. However, since so little is known about the natural longevity of individual amphibians (either xeric or mesic species) or about the ambient conditions prevailing over long periods, it is difficult to prove that a female may indeed get more than one opportunity to breed during her lifetime, even if she is potentially capable of doing so.

6.2.2.3
Population Ecology of Amphibians Living in Xeric Habitats

Population estimates are difficult to obtain for amphibians (Rose and Armentrout 1974; Warburg 1986a,b, 1994). Although it might be assumed that the sex ratio is normally 1:1, there are no reliable studies proving it. There are more males than females in populations of *Bufo americanum* (Christian and Taylor 1974). Husting (1965) found that the sex ratio in *Ambystoma maculatum* was biased in favour of males. The sex ratio in a xeric population of *S. s. infraimmaculata* was 1.4M:1F (Warburg 1986a) or almost 2M:1F in a mesic population (Joly 1959; Degani and Mendelssohn 1982). Likewise, in a population of *Ambystoma talpoideum*, the sex ratio shifted towards more males (Raymond and Hardy 1990). One possible reason could be the fact that males are more active, arrive earlier at the ponds than do females, and stay there longer (Degani and Warburg 1978). The sex ratio in the frog *Syrrhophus marnocki* was found to be 2M:1F (Jameson 1955).

Theoretically, such studies should be conducted on cohorts born to individual females. Since it is rather difficult to distinguish between the sexes from their external morphology at either the larval or the juvenile stages, it is necessary to wait until maturation in order to distinguish between the genders. As there is considerable mortality before reaching adulthood, it is virtually impossible to obtain an accurate estimate of the sex ratio of an entire cohort. The sex ratio of the adult population of a cohort is probably the outcome of differential survival. Males may die sooner than females since in some taxa sperm can be stored for one or more reproductive seasons. Consequently, the survival of males may not be selected. Nothing is known 'about these points.

There are several reports of anuran males competing for females (Balinsky and Balinsky 1954). This phenomenon has also been observed in urodeles (Kastle 1986). The reason for this phenomenon could be that in urodeles, the sperm of the male which mated last is responsible for the fertilization of the ova. Direct genetic proof is lacking, however. In no study were the offspring analysed genetically to see whether they genetically resembled the last mating male.

6.2.2.4
Longevity

Many toads fail to survive for more than a few years. Thus, 90–95% post-metamorphic juvenile *Bufo boreas* die during their first year (Samollow 1980). *Bufo gutturalis* (probably *B. regularis*) can survive for 2–3 years (Balinsky 1985). In *B. cognatus*, 2–5-year-old toads are sexually mature. Only about 11% of a batch survives to maturity (Blair 1953). In *B. valliceps*, inhabiting sub-xeric habitats in Texas, an even smaller percentage survives to maturity. Thus, about 6.2% females and 12.4% males survived their first year, with a mere 1.2% females and 0.6% males surviving until their fourth year (Blair 1960). About 25.8% adult mortality per breeding season was found in *Ambystoma talpoideum* (Raymond and Hardy 1990). Males of *A. maculatum* survive longer than females do (Husting 1965).

Over half the population of *B. punctatus* in Death Valley are first-year toads (Turner 1959). This means that many do not reach their maximal life expectancy of 4 years (Table 6.6). Longevity appears to be the key for survival in *B. punctatus* in the Colorado Desert since several batches of tadpoles are lost annually due to evaporation of the ponds before they can metamorphose. Since the adult life expectancy is 4 years, the female can perhaps afford to lose up to three batches of eggs, provided that breeding takes place every year and at least one batch survives to metamorphosis (Tevis 1966). *Cyclorana platycephalus* is capable of surviving 5 years of dormancy (Van Beurden 1980).

Other genera of frogs inhabiting xeric habitats, such as *Phrynobatrachus* and *Arthroleptis*, mature 4–5 months after metamorphosis (Barbault 1984a,b). In these frogs the life expectancy is low, but breeding is continuous. The salamander *S. s. infraimmaculata* has a life expectancy of over 20 years (Warburg 1994).

6.2.2.5
Density Effects

So far, there is no evidence that the density of the adult population affects the population size to any extent. This is in contrast to density effects on larval

Table 6.6. Life expectancy of amphibians from xeric habitats

Arthroleptis poecilonotus	4 months	Barbault and Trefaunt-Rodrigues (1979)
Bufo pentoni	6 years	Francillon et al. (1984)
B. cognatus	3–4 years	Blair (1953)
B. valliceps	Males 8 years; Females 5 years	Blair (1960)
Phrynobatrachus calcaratus	2 months	Barbault and Pilorge (1980)
Siren lacertina	3 years	Etheridge (1990b)

populations (discussed previously in this chapter). It does not mean, however, that no density-dependent competition exists among adult amphibians inhabiting xeric habitats. Since both food and suitable shelter may be scarce, it is possible that some such competition may, in fact, exist (Crump 1982). However, Kuzmin (1995) states that competition for food resources appears to be rare in amphibians under natural conditions.

6.2.2.6
Climatic Effects

Both larvae and adults are affected by climatic conditions which can also cause a delay in breeding, or even make breeding entirely impossible. We have seen that the larval amphibians are greatly affected by the unpredictable pattern of precipitation (see previous discussion). Drought may cause mortality in the adult population in spite of the fact that animals hide in suitable shelters and can survive for long periods without food. The emergence of aestivating amphibians is largely a response to temperature change and precipitation (Tester and Breckenridge 1964).

6.2.2.7
Population Cycles

There are many reports of amphibians disappearing from various regions in the world, including xeric habitats in remote areas. Since no accurate, long-term population census has ever been taken, there is no evidence to prove the existence of cyclic events in amphibian populations. There are, however, several short-term studies of amphibian populations in xeric habitats. In all of these, the population was studied over a limited period, and an estimate was made of the population's size. Such estimates depend to a large extent on the technique used for assessing the population. Even if the counts are made during periods of activity, there are constraints regarding the differential activity patterns of the two sexes which may affect the population count. Thus, in several amphibians living under xeric conditions, the males are more active than the females and stay abroad for longer periods. Consequently, they are likely to be encountered and counted more often than the females.

6.2.2.8
Population Explosion and Migration

In many species the simultaneous emergence of newly metamorphosed amphibians from a pond may give the false impression of a population explosion and subsequent migration taking place. However, this phenomenon lasts for only a short while before the juveniles disperse.

There is evidence in the literature for the existence of amphibian population explosions. The first and best known of these took place in biblical times. Since frogs were described there as coming out of the Nile, they could have been either a *Rana* species or perhaps *Bufo regularis*. Additional cases have been documented by Wager (1986), but these seem to fall under the category of migration since the movements were directional. Masses of *Cacosternum* sp. and *Brevipes* sp. moving on broad fronts have been recorded. In one instance, there was a report of large numbers of *Bufo carens* all migrating in the same direction. This mass movement lasted for 2 hours. Finally, Wager recorded large numbers of *Xenopus* sp. moving some distance from one pond to another.

6.2.2.9
Amphibian Communities

Little information is available about amphibian community structure (Woodward and Mitchell 1991). To what extent can groups of amphibians inhabiting the same hibitat be called communities, and to what extent are relationships shared within them? It seems as though large numbers just happen to be in the same place, occasionally sharing a breeding pond. Research on this subject is only just beginning. In deserts, amphibian communities mostly consist of anurans and involve only a few (2–5) species (Woodward and Mitchell 1991). Curiously, invertebrate taxa such as isopods and scorpions that inhabit deserts are likewise represented by only a few (2–5) species. Is this because deserts cannot support more than a few species of certain groups? Does this proportion hold for other regions as well? Much research is necessary before the nature of amphibian communities can be understood.

6.2.2.10
Conclusions

In conclusion, amphibians inhabiting xeric habitats have undergone ecological adaptations in the two main phases of their life histories: first, the immature, aquatic stage and second, the mature terrestrial stage. The aquatic amphibian is adapted to achieve metamorphosis as quickly as possible, at the largest possible size, and in sufficient numbers to ensure survival. Some species have reached a peak in rapid development (e.g. *Scaphiopus* spp.), whereas others metamorphosed at their maximal size (*Pelobates syriacus*).

The terrestrial stage has adapted in two ways, first in its readiness to reproduce whenever conditions are suitable climatically. This implies readiness of the gonads throughout much of the year. Second, the long life expectancy means that the risk of losing a cohort will be compensated by a second chance.

CHAPTER 7

Reflections and Conclusions

Scarcity of water has brought about a number of structural, behavioural, physiological and ecological adaptations in amphibians inhabiting seasonally xeric habitats. These have been discussed in previous chapters. I should now like to examine the information assembled here in an integrative way in order to determine whether new and fruitful avenues of research can be pointed out.

7.1
Habitat

There are rarely any studies in which attempts are made to define a xeric habitat. It is much easier to define an arid habitat, where the average annual rainfall is less than 50 mm, or a semi-arid one with not more than 250 mm rainfall. Any area with rainfall higher than this will be mesic. This definition is meaningless as regards amphibians since the way in which precipitation is divided seasonally, whether there are winter rains (or snow) and a dry summer, or alternatively if the summer is wet, are of such great significance.

Moreover, ambient data of temperature and humidity are generally based on meteorological measurements which are normally made in a Davidson Screen, 1–1.5 m above ground level. Again, such information is meaningless with respect to most amphibians, certainly to fossorial ones who dwell in the ground. Thus, basic microclimatic data are badly needed.

As we have seen, amphibians adapted to xeric conditions are found not only in arid and semi-arid habitats, but also in mesic ones that are seasonally dry. The majority of these amphibians are anurans, but representatives of three urodele families also inhabit xeric environments. Their adaptations to life under xeric conditions have evolved in various ways. The fact that they are found in such habitats indicates their ability to survive and reproduce success-fully under extremely harsh conditions.

Most of the research reported here has been conducted on representatives of two large amphibian faunas: the Australian and the North American. Others, Asian, African and South American, have been studied to a much lesser extent. This is true of studies on amphibians inhabiting both xeric and mesic environments, whether in their aquatic stages or as terrestrial adults.

7.2
Structure and Function

The main structures that enable amphibians to survive under xeric conditions involve the skin (epidermis and dermis), gills and lungs, kidney and urinary bladder, as well as the ovary. Each of these organs has undergone structural changes that enable the survival of the animal under such conditions. Whereas some of these structures have been described and their functions are known, in most cases the relationships between structure and function have been largely overlooked and thus are still very little understood.

The skin enables adult amphibians to reduce water loss partly by evolving lipid-secreting glands or, alternatively, by forming a cocoon. Moreover, water uptake through the skin takes place rapidly when water becomes available. Does the number of epidermal cell layers or the thickness of the skin affect evaporative water loss and uptake? We have no information about this.

Although a considerable amount of research deals with the transport of ions and the passage of water through the skin (dermis and epidermis), there is not a single study showing any special mechanism peculiar to amphibian species adapted to xeric conditions. It seems that all amphibians are capable of transporting ions and water through the skin. However, there is no explanation of the way in which amphibians from xeric habitats are capable of taking up water at a higher rate than those in a mesic one. Moreover, there does not seem to be any significant difference in water content between mesic and xeric amphibians. The differences in tolerance limits are not understood, nor is there any good explanation for such differences as far as they exist.

Structurally, the skin of immature, aquatic stages of amphibians differs greatly from that of the mature, terrestrial stages. This is true of all amphibians studied so far. No such difference has been found in skin structure between amphibians from various habitats.

Finally, a few species form cocoons during their aestivation. This structure seems to be the result of accumulating several layers of unsloughed stratum corneum. We have yet to find out how this is done. What could be stopping the corneal layer from sloughing off as is normally the case? What mechanism regulates cocoon formation, and how is it related to the lowered metabolism during aestivation?

Most water saving is due to kidney function via a decreased glomerular filtration rate controlled by neurohypophysial hormones. Another possible adaptation is the lower numbers of glomeruli in the kidney of amphibians from xeric habitats. Does the size and number of glomeruli affect kidney function? If so, in what way? Our knowledge in this field is still very limited.

We have previously seen that water can be saved by the excretion of uric acid as the end product of nitrogen metabolism – in other words, by shifting to uricotelism. This phenomenon has been studied in only a few anurans (rhacophorids, leptodactylids). Most of the amphibians living in xeric habitats are still ureotelic and conserve water in a different manner. As we have previously seen, the accumulation of urea is typical of those species that are not uricotelic (largely bufonids). The mechanism of this important adaptation is not clearly understood.

In some amphibian species (mainly bufonids, myobatrachids and pelobatids) the bladder functions as a water storage organ. What enables the bladder of one species to store water whereas in another it functions normally is not known. How does one explain the fact that water can be stored in the urinary bladder, the lymphatic system and coelomic cavity, as urine? After all, one might expect the normal physiological function to be the excretion of urine rather than its accumulation of it. Moreover, there does not seem to be any physiological damage as a result of such massive fluid accumulation. How can this be explained? There must be some endocrine control changing the normal function of the bladder to one of water storage. What could explain the fact that this control mechanism exists only in xeric-adapted species?

The difficulty in understanding the endocrine control of the water, ion and nitrogen balance of amphibians stems from the fact that they shift from aquatic to terrestrial life and vice versa. (We can see this during the return of newts to water to breed.) Thus, from being water conservers for part of the year, they become water eliminators since they have to get rid of excess water. Such a shift in function must also involve a major change in hormonal control, in hormone secretion, and in the receptors of the target organs (Warburg 1995). These fundamental problems have yet to be answered.

Finally, the ovarian structure, the number of oocytes, their size and stage of development are all reflected in the reproductive strategy of the species. Here we find special structural adaptations, as well as in the endocrine control of the ovarian cycle, that are peculiar to amphibians which are seasonally short of available water. Research in this field is badly needed, especially since it may shed some light on the reproductive strategies of endangered amphibian species.

7.3
Behavioural Responses

Exposure to high temperature is controlled mainly by a number of behavioural adaptations involving thermoregulation and the selection of optimal temperatures and moisture conditions. Most behavioural response studies have been conducted on the adult stage. Very few have been carried out on the larval stages. Here we are confronted with an array of conflicting data. Some may be due to the unknown physiological state of the experimental animals, others to the artificial conditions under which the experiments have been carried out. Ignoring ecological conditions may result in an erroneous understanding of behavioural patterns that appear to differ greatly from normal patterns; perhaps the experiments were conducted during different seasons or on different sexes or on females in different stages of their reproductive cycle.

There is no way in which circadian rhythms can be studied usefully in laboratory-reared animals since their normal pattern of rhythmical behaviour has been disrupted. They have been submitted to an abnormal light/dark regime, in addition to unnatural ambient conditions. In any case, only freshly collected animals should be used. Moreover, experiments conducted under artificial conditions are not necessarily related to rhythmic activity under natural conditions. It is essential that behavioural studies be conducted at least under semi-natural conditions in outdoor enclosures. Such studies are largely lacking. Based on the evidence presented here, we cannot conclude that amphibians adapted to life under xeric conditions show any specific behavioural adaptation that does not exist in other members of that class.

The basic physiology of water and thermal balance is perhaps well-known, but again in only a few selected amphibian species. Many of these studies have been conducted on animals of unknown origin (bought from a dealer). Consequently, neither the locality nor season of collecting, sometimes not even the sex or physiological conditions of the animals, are known to the researcher. It is essential to study all physiological parameters as soon as possible after an animal's capture. Furthermore, it is preferable that experimental animals be maintained under semi-natural conditions (perhaps in a suitable outdoor enclosure) during the course of any experiment.

Physiologists, including environmental physiologists, tend largely to disregard the ecological condition of the animals they are studying. At the same time, ecologists tend to ignore both the structures and the physiological functions of individual animals as if they had no role to play in the ecological interactions under investigation. Consequently, most researchers miss a great deal of important and relevant information that might have been available to them had they dissected a few of the specimens in their collections or evaluated their physiological and ecological conditions. This is of particular importance when studying reproductive strategy. In such studies, knowledge of the sex ratio of the population being investigated is of great importance, as well as knowledge of the breeding pattern. A glimpse at the ovary can tell a great deal about the previous and the forthcoming events in the reproductive life of an individual.

7.4
Ecological Adaptations

7.4.1
Breeding and Larval Stages

Present knowledge is largely based on field work. Consequently, there is as yet no information as to whether a female breeds only once during a season, or if she breeds at all, since an iteroparous species may skip a season if conditions are unsuitable for breeding. Long-term laboratory and field studies are badly needed to shed more light on this aspect.

Semelparity is extremely difficult to prove since it involves raising a female in isolation from birth in order to ensure her virginity. This female must then be mated and her reproductive cycle monitored until her death, in order to be certain that she did not breed more than once. Since all this can only be done in the laboratory, conditions are inevitably artificial and thus do not necessarily represent the natural situation.

The problems of reproducing under unpredictable weather conditions with insufficient precipitation have brought about a series of adaptations. First, rapid development of the larval stages up to metamorphosis may take place so that the juveniles can disperse early and seek shelter. On the other hand, the larval stages of species that have water available for breeding over longer periods can grow to a larger size. Consequently, the young metamorphs are also larger, a fact that has considerable advantages later in life.

The metamorphosing larvae of some species are capable of becoming cannibalistic, thereby gaining an extra food source which will enable them to grow even faster, become larger, and metamorphose sooner and at a greater size. All of these conditions are advantageous for an amphibian living under xeric conditions. Furthermore, since entire broods are likely to perish when pools of water dehydrate, some species became iteroparous, breeding more than once in their lifetime and thereby obtaining a second chance. Some of these species have a longer life expectancy and thus a better chance of producing at least one viable cohort. (A cohort is a brood of young originating from a single female and born on the same date.) On the other hand, due to multiple mating and sperm storage, they may have been fathered by different males. Genetic analysis is needed to confirm single parenthood. This type of study is still lacking.

There is considerable variability in the number of eggs in a batch. In the studies that are now available, scarcely any consideration has been given to the female's weight, and the fact that egg numbers are positively related to it is not taken into account. Data on egg diameter are largely missing. This parameter is variable and can change even while measurements are being taken. The course of vitellogenesis has been followed only in a few species, and data on this subject are badly needed. Likewise, the study of development time has been based mostly on laboratory studies. Only rarely have studies been carried out in artificial ponds. In these studies, it is essential to take into account food quantity and quality, density effects, and ambient conditions (temperature, pH, ions).

7.4.2
Longevity

Most work on population structure, sex ratio, adult competition, and community structure are based on field studies. Since methods vary considerably, it is difficult to draw any conclusions. On the other hand, in all except one study on *S. s. infraimmaculata*, longevity estimates have been based on laboratory observations. It is essential that animals of known age be released back into their habitat and their lives monitored over long periods of time. Skeletochronological studies have their limits. An annual ring is not always clear enough to be counted. Thus, estimates of the length of life are most probably conservative. A long life expectancy is of great importance to amphibians adapted to xeric conditions, who may obtain a second chance of breeding should the first attempt be unsuccessful.

To conclude, the adaptation of amphibians to more extreme terrestrial conditions was achieved in various ways involving modified structures and function, behaviour and ecology.

References

Accordi F, Grassi Milano E (1990) Evolution and development of the adrenal gland in amphibians. Fortschr Zool 38:257–268

Aceves J, Erlij D, Whittembury G (1970) The role of the urinary bladder in water balance of *Ambystoma mexicanum*. Comp Biochem Physiol 33:39–42

Adler K (1970) The role of extraoptic photoreceptors in amphibian rhythms and orientation: a review. J Herpetol 4:99–112

Adler K (1971) Pineal end organ: role in extraoptic entrainment of circadian locomotor rhythm in frogs. In: Menaker M (ed) Biochronometry. Proc Symp Natl Acad Sci, Friday Harbor, Washington, pp 342–350

Adler K, Taylor DH (1981) Toad orientation: variability of response and its relationship to individuality and environmental parameters. J Comp Physion [A] 144:45–51

Alford RA (1986) Habitat use and positional behavior of anuran larvae in a northern Florida temporary pond. Copeia 1986:408–423

Alford RA, Harris RN (1988) Effects of larval growth history on anuran metamorphosis. Am Nat 131:91–106

Alvarado RH (1967) The significance of grouping on water conservation in *Ambystoma*. Copeia 1967:667–668

Alvarado RH (1972) The effects of dehydration on water and electrolytes in *Ambystoma tigrinum*. Physiol Zool 45:43–53

Alvarado RH (1979) Amphibians. In: Maloyi GMO (ed) Comparative physiology of osmoregulation in animals, vol 1. Academic Press, London, pp 261–303

Alvarado RH, Johnson SR (1965) The effects of arginine vasotocin and oxytocin on sodium and water balance in *Ambystoma*. Comp Biochem Physiol 16:531–546

Alvarado RH, Johnson SR (1966) The effects of neurohypophysial hormones on water and sodium balance in larval and adult bullfrogs (*Rana catesbeiana*). Comp Biochem Physiol 18:549–561

Alvarado RH, Kirschner LB (1963) Osmotic and ionic regulation in *Ambystoma tigrinum*. Comp Biochem Physiol 10:55–67

Alvarado RH, Kirschner LB (1964) Effect of aldosterone on sodium fluxes in larval *Ambystoma tigrinum*. Nature 202:922–923

Amey AP, Grigg GC (1995) Lipid-reduced evaporative water loss in two arboreal hylid frogs. Comp Biochem Physiol [A] 111:283–291

Anderson JD (1967) A comparison of the life histories of coastal and montane populations of *Ambystoma macrodactylum* in California. Am Midl Nat 77:323–355

Anderson JD (1972) Phototactic behavior of larvae and adults of two subspecies of *Ambystoma macrodactylum*. Herpetologica 28:222–226

Anderson JD, Graham RE (1967) Vertical migration and stratification of larval *Ambystoma*. Copeia 1967:371–374

Anderson JD, Webb RG (1978) Life history aspects of the Mexican salamander *Ambystoma rosaceum* (Amphibia, Urodela, Ambystomatidae). J Herpetol 12:89–93

Anderson JD, Williamson GK (1974) Nocturnal stratification in larvae of the mole salamander, *Ambystoma talpoideum*. Herpetologica 30:28–29

Anderson JD, Worthington RD (1971) The life history of the Mexican salamander *Ambystoma ordinarium* Taylor. Herpetologica 27:165–176

Anderson JD, Hassinger DD, Dalrymple GH (1971) Natural mortality of eggs and larvae of *Ambystoma t. tigrinum*. Ecology 52:1107–1112

Andren C, Nilson G (1979) A new species of toad (Amphibia, Anura, Bufonidae) from the Kavir desert, Iran. J Herpetol 13:93–100

Bagnara JT, Ferandez PJ (1993) Hormonal influences on the development of amphibian pigmentation patterns. Zool Sci 10:733–748

Bailey WJ, Roberts JD (1981) The bioacoustics of the burrowing frog *Heleioporus* (Leptodactylidae). J Nat Hist 15:693–702

Bakker HR, Bradshaw SD (1977) Effect of hypothalamic lesions on water metabolism of the toad *Bufo marinus*. J Endocrinol 75:161–172

Baldwin GF, Bentley PJ (1981a) A role of skin in Ca metabolism of frog? Comp Biochem Physiol [A] 68:181–185

Baldwin GF, Bentley PJ (1981b) Calcium exchanges in two neotenic urodeles: *Necturus maculosus* and *Ambystoma tigrinum*. Role of the integument. Comp Biochem Physiol [A] 70:65–68

Baldwin GF, Bentley PJ (1982) Roles of the skin and gills in sodium and water exchanges in neotenic urodele amphibians. Am J Physiol 242:R94–R96

Baldwin RA (1974) The water balance response of the pelvic "patch" of *Bufo punctatus* and *Bufo boreas*. Comp Biochem Physiol [A] 47:1285–1295

Balinsky BI (1957) South African Amphibia as material for biological research. S Afr J Sci 53:383–391

Balinsky BI (1962) Patterns of animal distribution on the African continent. Ann Cape Provis Mus II:2299–310

Balinsky BI (1969) The reproductive ecology of amphibians of the Transvaal highveld. Zool Afr 4:37–93

Balinsky BI (1985) Observations on the breeding of toads in a restricted habitat. S Afr J Zool 20:61–64

Balinsk BI, Balinsky JB (1954) On the breeding habits of the South African bullfrog, *Pyxicephalus adspersus*. S Afr J Sci 51:55–58

Balinsky JB (1970) Nitrogen metabolism in amphibians. In: Campbell JW (ed) Comparative biochemistry of nitrogen metabolism, vol 2. The vertebrates. Academic Press, London, pp 519–537

Balinsky JB (1981) Adaptation of nitrogen metabolism to hyperosmotic environment in Amphibia. J Exp Zool 215:335–350

Balinsky JB, Baldwin E (1961) The mode of excretion of ammonia and urea in *Xenopus laevis*. J Exp Biol 38:695–705

Balinsky JB, Cragg MM, Baldwin E (1961) The adaptation of amphibian waste nitrogen excretion to dehydration. Comp Biochem Physiol 3:236–244

Balinsky JB, Choritz EL, Coe CGL, van der Schans GS (1967) Amino acid metabolism and urea excretion in naturally aestivating *Xenopus laevis*. Comp Biochem Physiol 22:59–68

Balinsky JB, Chemaly SM, Currin AE, Lee AR, Thompson RL, van der Westhuizen DR (1976) A comparative study of enzymes of urea and uric acid metabolism in different species of Amphibia, and the adaptation to the environment of the tree frog *Chiromantis xerampelina* Peters. Comp Biochem Physiol [B] 54:549–555

Ball RW, Jameson DL (1970) Biosystematics of the canyon tree frog *Hyla cadaverina* Cope (=*Hyla californiae* Gorman). Proc Calif Acad Sci 37:363–380

Ballinger RE, McKinney CO (1966) Developmental temperature tolerance of certain anuran species. J Exp Zool 161:21–28

Bani G, Cecchi R, Bianchi S (1985) Skin morphology in some amphibians with different ecological habits. Z Mikrosk Anat Forsch 99:455–474

Barbault R (1984a) Pression de prédation et évolution des stratégies démographiques en zone tropicale: le cas des lezards et des amphibiens. Rev Zool Afr 101:301–327

Barbault R (1984b) Strategies de reproduction et démographie de quelques amphibiens anoures tropicaux. Oikos 43:77–87

Barbault R, Pilorge T (1980) Observations sur la reproduction et la dynamique des populations de quelques anoures tropicaux. V. *Phrynobatrachus calcaratus*. Acta Oecol Gen 1:373–382

Barbault R, Trefaut-Rodrigues M (1978) Observations sur la reproduction et la dynamique des populations de quelques anoures tropicaux. I. *Ptychadena maccarthyensis* et *Ptychadena oxyrhynchus*. Terre Vie 32:441–452

Barbault R, Trefaut-Rodrigues M (1979) Observations sur la réproduction et la dynamique des populations de quelques anoures tropicaux. III. *Arthroleptis poecilonotus*. Trop Ecol 20:64–77

Becker HE, Cone RA (1966) Light-stimulated electrical responses from skin. Science 154:1051–1053

Beiswenger RE (1978) Responses of *Bufo* tadpoles (Amphibia, Anura, Bufonidae) to laboratory gradients of temperature. J Herpetol 12:499–504

Belehradek J, Huxley JS (1928) The effects of pituitrin and of narcosis on water-regulation in larval and metamorphosed *Amblystoma*. J Exp Biol 5:89–96

Beneski JT, Zalisko EJ, Larsen JH (1986) Demography and migratory patterns of the eastern long-toed salamander, *Ambystoma macrodactylum columbianum*. Copeia 1986:398–408

Bennett AF (1978) Activity metabolism of the lower vertebrates. Annu Rev Physiol 40:447–469

Bentley PJ (1958) The effects of neurohypophysial extracts on water transfer across the wall of the isolated urinary bladder of the toad *Bufo marinus*. J Endocrinol 17:201–209

Bentley PJ (1959) The effects of neurohypophysial extracts on the tadpole of the frog, *Heleioporus eyrei*. Endocrinology 64:69

Bentley PJ (1966a) Adaptations of Amphibia to arid environments. Science 152:619–623

Bentley PJ (1966b) The physiology of the urinary bladder of Amphibia. Biol Rev 41:275–316

Bentley PJ (1967a) Natriferic and hydro-osmotic effects on the toad bladder of vasopressin analogues with selective antidiuretic activity. J Endocrinol 39:299–304

Bentley PJ (1967b) The role of the toad urinary bladder in the amphibian "water balance effect" of neurohypohysial hormones. J Endocrinol 37:349–350

Bentley PJ (1969) Neurohypophyseal hormones in Amphibia: a comparison of their actions and storage. Gen Comp Endocrinol 13:39–44

Bentley PJ (1971) Sodium and water movement across the urinary bladder of a urodele amphibian (the mudpuppy *Necturus maculosus*): studies with vasotocin and aldosterone. Gen Comp Endocrinol 16:356–362

Bentley PJ (1974) Actions of neurohypophysial peptides in amphibians, reptiles and birds. In: Greep RO, Astwood EB (eds) Handbook of physiology. Endocrinology, vol IV, part 2. American Physiological Society, Bethesda, pp 545–563

Bentley PJ (1983) Urinary loss of calcium in an anuran amphibian (*Bufo marinus*) with a note on the effects of calcemic hormones. Comp Biochem Physiol [B] 76:717–719

Bentley PJ (1984) Calcium metabolism in the Amphibia. Comp Biochem Physiol [A] 79:1–5

Bentley PJ (1987) Actions of hormones on salt and water transport across cutaneous and urinary bladder epithelia. In: Pang PKT, Schreibman MP (eds) Vertebrate endocrinology: fundamentals and biomedical implications, vol 2. Regulation of water and electrolytes. Academic Press, New York, pp 271–291

Bentley PJ, Baldwin GF (1980) Comparison of transcutaneous permeability in skins of larval and adult salamanders (*Ambystoma tigrinum*). Am J Physiol 239:R505–R508

Bentley PJ, Greenwald L (1970) Neurohypophysial function in bullfrog (*Rana catesbeiana*) tadpoles. Gen Comp Endocrinol 14:412–415

Bentley PJ, Heller H (1964) The action of neurohypophysial hormones on the water and sodium metabolism of urodele amphibians. J Physiol (Lond) 171:434–453

Bentley PJ, Heller H (1965) The water-retaining action of vasotocin on the fire salamander (*Salamandra maculosa*): the role of the urinary bladder. J Physiol (Lond) 181:124–129

Bentley PJ, Main AR (1972) Zonal differences in permeability of the skin of some anuran Amphibia. Am J Physiol 223:361–363

Bentley PJ, Yorio T (1976) The passive permeability of the skin of anuran Amphibia: a comparison of frogs (*Rana pipiens*) and toads (*Bufo marinus*). J Physiol (Lond) 261:603–615

Bentley PJ, Yorio T (1979) Evaporative water loss in anuran Amphibia: a comparative study. Comp Biochem Physiol [A] 62:1005–1009

Bentley PJ, Lee AK, Main AR (1958) Comparison of dehydration and hydration of two genera of frogs (*Heleioporus* and *Neobatrachus*) that live in areas of varying aridity. J Exp Biol 35:677–684

Bern HA (1975) Prolactin and osmoregulation. Am Zool 15:937–948

Bhaduri JL, Basu SL (1957) A study of the urinogenital system of Salientia. I. Ranidae and Hyperoliidae of Africa. Ann Mus R Congo Belge Tervuren S 8 Sci Zool 56 pp

Bieniak A, Watka R (1962) Vascularization of respiratory surfaces in *Bufo cognatus* Say and *Bufo compactilis* Wiegman. Bull Acad Pol Sci 10:9–12

Binkley S (1979) Pineal rhythms in vivo and in vitro. Comp Biochem Physiol [A] 64:201–206

Birukow G (1950) Vergleichende Untersuchungen über das Helligkeits- und Farbensehen bei Amphibien. Z Vergl Physiol 32:348–382

Bizer JR (1978) Growth rates and size at metamorphosis of high elevation populations of *Ambystoma tigrinum*. Oecologia 34:175–184

Blackburn DG (1994) Discrepant usage of the term "ovoviviparity" in the herpetological literature. Herpetol J 4:65–72

Blair WF (1953) Growth, dispersal and age at sexual maturity of the Mexican toad (*Bufo valliceps*, Wiegmann). Copeia 1953:208–212

Blair WF (1960) A breeding population of the Mexican toad (*Bufo valliceps*) in relation to its environment. Ecology 41:165–174

Blair WF (1961) Calling and spawning seasons in a mixed population of anurans. Ecology 42:99–110

Blair WF (1975) Adaptation of anurans to equivalent desert scrub of North and South America. In: Goodall DW (ed) Evolution of desert biota. University of Texas Press, Austin, pp 197–222

Blaustein AR, Waldman B (1992) Kin recognition in anuran amphibians. Anim Behav 44:207–221

Blaustein AR, Walls SC (1995) Aggregation and kin recognition. In: Heatwole H (ed) Amphibian biology, vol 2. Social behaviour. Surrey Beatty, Sydney, pp 569–602

Blaustein AR, Yoshikawa T, Asoh K, Walls SC (1993) Ontogenetic shifts in tadpole kin recognition: loss of signal and perception. Anim Behav 46:525–538

Blaylock LA, Ruibal R, Platt-Aloia K (1976) Skin structure and wiping behavior of Phyllomedusine frogs. Copeia 1976:283–295

Bobrov VV (1986) Zur zoogeographischen Analyse der Herpetofauna der Mongolei. In: Herpetologische Untersuchungen in der Mongolischen Volksrepublik. Soviet Academy of Sciences, Moscow, pp 85–95 (in Russian)

Boell EJ, Greenfield P, Hille B (1963) The respiratory function of gills in the larvae of *Amblystoma punctatum*. Dev Biol 7:420–431

Boernke WE (1973) Adaptations of amphibian arginase. I. Response to dehydration. Comp Biochem Physiol [B] 44:647–655

Boernke WE (1974) Natural variations in hepatic and kidney arginase activities in Minnesota anuran amphibians. Comp Biochem Physiol [B] 47:201–207

Boisseau C, Joly J (1975) Transport and survival of spermatozoa in female Amphibia. In: Hafez ESE, Thibault CG (eds) The biology of spermatozoon. S Karger, Basel, pp 94–104

Bolton PM, Beuchat CA (1991) Cilia in the urinary bladder of reptiles and amphibians: a correlate of urate production. Copeia 1991:711–717

Bond AN (1960) An analysis of the response of salamander gills to changes in the oxygen concentration of the medium. Dev Biol 1:1–20

Boschwitz D (1967) On the problem of seasonal changes in anuran parathyroids: observations on *Bufo viridis* in Israel. Isr J Zool 16:46–48

Boschwitz D (1977) Seasonal changes in the histological structure of the ultimobranchial body of the toad *Bufo viridis*. Br J Herpetol 5:719–726

Boutilier RG, Randall DJ, Shelton G, Toews DP (1979a) Acid-base relationships in the blood of the toad, *Bufo marinus*. II. The effects of dehydration. J Exp Biol 82:345–355

Boutilier RG, Randall DJ, Shelton G, Toews DP (1979b) Acid-base relationships in the blood of the toad, *Bufo marinus*. III. The effects of burrowing. J Exp Biol 82:357–365

Bowker RG, Bowker MH (1979) Abundance and distribution of anurans in a Kenyan pond. Copeia 1979:278–285

Boyd SK (1992) Sexual mechanisms in hormonal control of release calls in bullfrog. Horm Behav 26:522–535

Boyd SK (1994a) Arginine vasotocin facilitation of advertisement calling and call phonotaxis in bullfrogs. Horm Behav 28:232–240

Boyd SK (1994b) Development of vasotocin pathways in the bullfrog brain. Cell Tissue Res 276:593–602

Boyd SK, Tyler CJ, De Vries GJ (1992) Sexual dimorphism in the vasotocin system of the bullfrog (*Rana catesbeiana*). J Comp Neurol 325:313–325

Bragg AW (1944) The spadefoot toads in Oklahoma with a summary of our knowledge of the group I. Am Nat 78:517–533

Bragg AN (1945) The spadefoot toads in Oklahoma with a summary of our knowledge of the group II. Am Nat 79:52–72

Bragg AN (1954) Aggregational behavior and feeding reactions in tadpoles of the Savannah spadefoot. Herpetologica 10:97–102

Bragg AN (1956) Dimorphism and cannibalism in tadpoles of *Scaphiopus bombifrons* (Amphibia, Salientia). Southwest Nat 1:105–108

Bragg AN (1959) Behavior of tadpoles of Hurter's spadefoot during an exceptionally rainy season. Wasmann J Biol 17:23–42

Bragg AN (1960) Experimental observations of the feeding of spadefoot tadpoles. Southwest Nat 5:201–207

Bragg AN (1961) A theory of the origin of spade-footed toads deduced principally by a study of their habits. Anim Behav 9:178–186

Bragg AN (1962) Predator-prey relationship in two species of spadefoot tadpoles with notes on some other features of their behavior. Wassmann J Biol 20:81–97

Bragg AN (1965) Gnomes of the night. University of Pennsylvania Press, Philadelphia, 127 pp

Bragg AN, Brook M (1958) Social behavior in juveniles of *Bufo cognatus* Say. Herpetologica 14:141–147

Bragg AN, King OM (1960) Aggregational and associated behavior in tadpoles of the plains spadefoot. Wasmann J Biol 18:273–289

Bragg AN, Smith CC (1943) Observations on the ecology and natural history of Anura. IV. The ecological distribution of toads in Oklahoma. Ecology 24:285–309

Brandon RA (1976) Spontaneous and induced metamorphosis of *Ambystoma dumerilii* (Duges), a paedogenetic Mexican salamander, under laboratory conditions. Herpetologica 32:429–438

Brandon RA (1989) Natural history of the axolotl and its relationship to other ambystomatid salamanders. In: Armstrong JB, Malacinski GM (eds) Developmental biology of the axolotl. Oxford University Press, New York, pp 13–21

Brattstrom BH (1962) Homing in the giant toad, *Bufo marinus*. Herpetologica 18:176–180

Brattstrom BH (1963) A preliminary review of the thermal requirements of amphibians. Ecology 44:238–254

Brattstrom BH (1968) Thermal acclimation in anuran amphibians as a function of latitude and altitude. Comp Biochem Physiol 24:93–111

Brattstrom BH (1970a) Thermal acclimation in Australian amphibians. Comp Biochem Physiol 35:69–103

Brattstrom BH (1970b) Amphibia. In: Whittow GC (ed) Comparative physiology of thermoregulation. Academic Press, New York, pp 135–166

Brattstrom BH (1979) Amphibian temperature regulation studies in the field and laboratory. Am Zool 19:345–356

Brattstrom BH, Lawrence P (1962) The rate of thermal acclimation in anuran amphibians. Physiol Zool 35:148–156

Braun EJ, Dantzler WH (1987) Mechanisms of hormone actions on renal function. In: Pang PKT, Schreibman MP (eds) Vertebrate endocrinology: fundamentals and biomedical implications, vol 2. Water and electrolytes. Academic Press, New York, pp 189–210

Brekke DR, Hillyard SD, Winokur RM (1991) Behavior associated with the water absorption response by the toad, *Bufo punctatus*. Copeia 1991:393–401

Brewer KJ, Hoytt BJ, McKeown BA (1980) The effects of prolactin, corticosterone and ergocryptine on sodium balance in the urodele, *Ambystoma gracile*. Comp Biochem Physiol [C] 66:203–208

Brizzi R, Delfino G, Selmi MG, Sever DM (1995) Spermathecae of *Salamandrina terdigitata* (Amphibia: Salamandridae): patterns of sperm storage and degredation. J Morphol 223:21–33

Brockelman WY (1969) An analysis of density effects and predation in *Bufo americanus* tadpoles. Ecology 50:632–644

Brown HA (1967) High temperature tolerance of the eggs of a desert anuran, *Scaphiopus hammondi*. Copeia 1967:365–370

Brown HA (1969) The heat resistance of some anuran tadpoles (Hylidae and Pelobatidae). Copeia 1969:138–147

Brown PS, Brown SC (1987) Osmoregulatory actions of prolactin and other adenohypophysial hormones. In: Pang PKT, Schreibman MP (eds) Vertebrate endocrinology: fundamentals and biomedical implications, vol 2. Regulation of water and electrolytes. Academic Press, New York, pp 45–84

Brown PS, Brown SC, Bisceglio IT, Lemke SM (1983) Breeding conditions, temperature, and the regulation of salt and water by pituitary hormones in the red-spotted newt, *Notophthalmus viridescens*. Gen Comp Endocrinol 51:292–302

Brown PS, Barry B, Brown SC (1988) Changes in absolute and proportional water content during growth and metamorphosis of *Rana sylvatica*. Comp Biochem Physiol [A] 91:189–194

Brown SC, Horgan EA, Savage LM, Brown PS (1986) Changes in body water and plasma constituent during bullfrog development: effects of temperature and hormones. J Exp Zool 237:25–33

Brzoska J, Schneider H (1982) Territorial behavior and vocal response in male *Hyla arborea savignyi* (Amphibia: Anura). Isr J Zool 31:27–37

Budtz PE (1977) Aspects of moulting in anurans and its control. Symp Zool Soc Lond 39:317–334

Budtz PO, Larsen LO (1973) Structure of the toad epidermis during the moulting cycle. I. Light microscopic observations in *Bufo bufo* (L.). Z Zellforsch 144:353–368

Bundy D, Tracy CR (1977) Behavioral response of American toads (*Bufo americanus*) to stressful thermal and hydric environments. Herpetologica 33:455–458

Burggren WW, Wood SC (1981) Respiration and acid-base balance in the salamander, *Ambystoma tigrinum*: influence of temperature acclimation and metamorphosis. J Comp Physiol [B] 114:241–246

Butman BT, Obika M, Tchen TT, Taylor JD (1979) Hormone-induced pigment translocations in amphibian dermal iridiophores, in vitro: changes in cell shape. J Exp Zool 208:17–34

Buttemer WA (1990) Effect of temperature on evaporative water loss of the Australian tree frogs *Litoria caerulea* and *Litoria chloris*. Physiol Zool 63:1043–1057

Bytinsky-Salz H (1976) Structural alterations in the epidermis and especially the coverplate in anuran tadpoles after osmotic changes. In: Proc 6th Eur Congr on Electron microscopy. Jerusalem, pp 574–576

Campbell JW, Vorhaben JE, Smith DD (1987) Uricotely: its nature and origin during evolution of tetrapod vertebrates. J Exp Zool 243:349–363

Campantico E, Dore B, Guardabassi A, Guastalla A, Mastracchio G (1974) Influence of environmental changes on blood calcium level and $CaCO_3$ deposits in anuran amphibians treated or not with prolactin. Boll Zool 41:129–138

Campantico E, Guastalla A, Patriarca E (1985) Identification of immunofluorescence of prolactin- and somatotrophin-producing cells in the pituitary gland of the tree frog *Hyla arborea*. Gen Comp Endocrinol 57:110–116

Canziani GA, Cannata MA (1980) Water balance in *Ceratophrys ornata* from two different environments. Comp Biochem Physiol [A] 66:599–603

Carey C (1978) Factors affecting body temperatures of toads. Oecologia 35:197–219

Carey C (1979a) Effect of constant and fluctuating temperatures on resting and active oxygen consumption of toads, *Bufo boreas*. Oecologia 39:201–212

Carey C (1979b) Aerobic and anaerobic energy expenditure during rest and activity in Montane *Bufo b. boreas* and *Rana pipiens*. Oecologia 39:213–228

Carlisky NJ, Botbol V, Barrio A, Sandik LI (1968) Renal handling of urea in three preferentially terrestrial species of amphibian Anura. Comp Biochem Physiol 26:573–578

Carr JA, Norris DO (1988) Interrenal activity during metamorphosis of the tiger salamander, *Ambystoma tigrinum*. Gen Comp Endocrinol 71:63–69

Carter DB (1979) Structure and function of the subcutaneous lymph sacs in the Anura (Amphibia). J Herpetol 13:321–327

Castañe PM, Salibian A, Zylbersztein C, Herkovits J (1987) Ontogenic screening of aldosterone in the South American toad *Bufo arenarum* (Hensel). Comp Biochem Physiol [A] 86:697–701

Castañe PM, Rovedatti MG, Salibian A (1990) Liver arginase activity and urinary nitrogen products profile in adult *Bufo arenarum* under different protein intake. Life Sci 46:1893–1901

Cei JM (1955) Chacoan batrachians in Central Argentina. Copeia 1955:291–293

Cei JM (1959) Ecological and physiological observations on polymorphic populations of the toad *Bufo arenarum* Hensel, from Argentine. Evolution 13:532–536

Cei JM (1962) Batracios de Chile. University of Chile, Santiago de Chile, 128 pp

Cei JM (1980) Amphibians of Argentina. Monit Zool Ital Monogr 2:1–609

Channing A (1976) Life histories of frogs in the Namib desert. Zool Afr 11:299–312

Chapman A, Dell J (1985) Biology and zoogeography of the amphibians and reptiles of the Western Australian wheatbelt. Rec West Aust Mus 12:1–46

Chapman BM, Chapman RF (1957) Field study of a population of leopard toads (*Bufo regularis regularis*). J Anim Ecol 27:265–286

Chauvet J, Ouedraogo Y, Michel G, Acher R (1993) Vasotocin and hydrin 2 (vasotocinyl-gly) in the African toad *Bufo regularis*: study under various environmental conditions. Comp Biochem Physiol [A] 104:497–502

Chiba A, Kikucki M, Aoki K (1995) Entrainment of the circadian locomotor activity rhythm in the Japanese newt by melatonin injections. J Comp Physiol [A] 176:473–477

Christian D, Taylor DH (1978) Population dynamics in breeding aggregations of the American toad, *Bufo americanus* (Amphibia, Anura, Bufonidae). J Herpetol 12:17–24

Christian K, Green B (1994) Water flux in the Australian treefrog *Litoria caerulea* under natural and semi-natural conditions. Amphibia-Reptilia 15:401–405

Christensen CU (1974) Adaptations in the water economy of some anuran Amphibia. Comp Biochem Physiol [A] 47:1035–1049

Christensen CU (1975) Effects of dehydration, vasotocin and hypertonicity on net water flux through the isolated, perfused pelvic skin of *Bufo bufo bufo* (L.). Comp Biochem Physiol [A] 51:7–10

Cirne BR, Reis HA, Silveira JEN (1981) Skin water uptake and renal function in the toad. Comp Biochem Physiol [A] 69:219–224

Claussen DL (1969) Studies on water loss and rehydration in anurans. Physiol Zool 42:1–14

Claussen DL (1977) Thermal acclimation in ambystomatid salamanders. Comp Biochem Physiol [A] 58:333–340

Cloudsley-Thompson JL (1967) Diurnal rhythm, temperature and water relations of the African toad, *Bufo regularis*. J Zool (Lond) 152:43–54

Cloudsley-Thompson JL (1970) The significance of cutaneous respiration in *Bufo regularis* Reuss. Int J Biometereol 14:361–364

Cloudsley-Thompson JL (1974) Water relations of the African toad, *Bufo mauritanicus* Schl. Br J Herpetol 5:425–426

Cochran DM (1961) Amphibians of the world. Hamish Hamilton, London, 199 pp

Coe M (1974) Observations on the ecology and breeding biology of the genus *Chiromantis* (Amphibia: Rhacophoridae). J Zool (Lond) 172:13–34

Cogger HG (1975) Reptiles and amphibians of Australia. Reed Books, Chastwood, New South Wales

Cogger HG (1992) Reptiles and amphibian of Australia, revised edn. Reed Books, Chastwood, New South Wales, 584 pp

Coleman R (1975) The development and fine structure of ultimobranchial glands in larval anurans. II. *Bufo viridis, Hyla arborea, Rana ridibunda*. Cell Tissue Res 164:215–232

Collett TS (1982) Do toads plan routes? A study of the detour behaviour of *Bufo viridis*. J Comp Physiol [A] 146:261–271

Collins JP (1979) Sexually mature larvae of the salamanders *Ambystoma rosaceum* and *A. tigrinum velasci* from Chihuahua, Mexico: taxonomic and ecologic notes. J Herpetol 13:351–354

Collins JP (1981) Distribution, habitats and life history variation in the tiger salamander, *Ambystoma tigrinum*, in East-Central and Southwest Arizona. Copeia 1981:666–675

Collins JP, Cheek JE (1983) Effect of food and density on development of typical and cannibalistic salamander larvae in *Ambystoma tigrinum nebulosum*. Am Zool 23:77–84

Collins JP, Holomuzki JR (1984) Intraspecific variation in diet within and between trophic morphs in larval tiger salamanders (*Ambystoma tigrinum nebulosum*). Can J Zool 62:168–174

Conant R (1977) Semiaquatic reptiles and amphibians of the Chihuahuan Desert and their relationships to drainage patterns of the region. In: Wauer RH, Riskind DH (eds) Transactions of the Symposion on the Biological Resources of the Chihuauan Desert region. US Dept Interior, National Park Service transactions, vol 3, pp 455–491

Cortelyou JR (1967) The effect of commercial prepared parathyroid extracts on plasma and urine calcium levels in *Rana pipiens*. Gen Comp Endocrinol 9:234–240

Cortelyou JR, McWhinnie DJ (1967) Parathyroid glands of amphibians. I. Parathyroid structure and function in the Amphibia, with emphasis on regulation of mineral ions in body fluids. Am Zool 7:843–855

Coviello A (1969) Tubular effect of angiotensin II on the tadpole kidney. Acta Physiol Lat Am 19:73–82

Coviello A (1970) Hydrosmotic effect of angiotensin II: isolated toad bladder. Acta Physiol Lat Am 22:218–226

Cowles EH, Bogert CM (1944) A preliminary study of the thermal requirements of desert reptiles. Bull Am Mus Nat Hist 83:261–296

Crabbe J (1961) Stimulation of active sodium transport across the isolated toad bladder after injection of aldosterone to the animal. Endocrinology 69:673–682

Cragg MM, Balinsky JB, Baldwin E (1961) A comparative study of nitrogen excretion in some Amphibia and reptiles. Comp Biochem Physiol 3:227–235

Crews D, Silver R (1985) Reproductive physiology and behavior interactions in nonmammalian vertebrates. In: Adler N, Pfaff D, Goy RW (eds) Handbook of behavioral neurobiology, vol 7. Plenum Press, New York, pp 101–182

Creusere FM, Whitford WG (1976) Ecological relationships in a desert anuran community. Herpetologica 32:7–18

Crocker CE, Stiffler DF (1991) The effects of amiloride on electrolyte transport and acid-base balance in the larval salamander, *Ambystoma tigrinum*. J Comp Physiol [B] 161:460–464

Crump ML (1982) Amphibian reproductive ecology on the community level. US Fish Wildl Serv Wildl Res Rep 13:21–36

Crump ML (1983) Opportunistic cannibalism by amphibian larvae in temporary aquatic environment. Am Nat 121:281–287

Crump ML (1986) Cannibalism by younger tadpoles: another hazard of metamorphosis. Copeia 1986:1007–1009

Crump ML (1989) Effect of habitat drying on developmental time and size at metamorphosis in *Hyla pseudopuma*. Copeia 1989:794–797

Crump ML (1990) Possible enhancement of growth in tadpoles through cannibalism. Copeia 1990:560–564

Crump ML (1992) Cannibalism in amphibians. In: Elgar MA, Crespi BJ (eds) Cannibalism. Oxford University Press, Oxford, pp 256–276

Cutz E, Goniakowska-Witalinska L, Chan W (1986) An immunohistochemical study of regulatory peptides in lungs of amphibians. Cell Tissue Res 244:227–233

Czopek G, Czopek J (1959) Vascularization of respiratory surfaces in *Bufo viridis* Laur. and *Bufo calamita* Laur. Bull Acad Pol Sci 7:39–45

Czopek J (1955) The vascularization of the respiratory surfaces of some Salientia. Zool Pol 6:101–134

Czopek J (1957) The vascularization of respiratory surfaces in *Ambystoma mexicanum* (Cope) in ontogeny. Zool Pol 8:131–149

Czopek J (1959) Vascularization of respiratory surfaces in *Salamandra salamandra* L. in ontogeny. Bull Acad Pol Sci 7:473–478

Czopek J (1962) Vascularization of respiratory surfaces in some Caudata. Copeia 1962:576–587

Czopek J, Rys T, Szemro B (1968) The distribution of capillaries in the respiratory surfaces of *Scaphiopus holbrooki* and *Scaphiopus couchi* Baird. Zool Pol 18:117–124

Dash MC, Hota AK (1980) Density effects on the survival, growth rate, and metamorphosis of *Rana tigrina* tadpoles. Ecology 61:1025–1028

Dash MC, Mahanta JK (1993) Quantitative analysis of the community structure of tropical amphibian assemblages and its significance to conservation. J Biosci 18:121–139

Davis WL, Goodman DBP (1986) Antidiuretic hormone response in the amphibian urinary bladder: time course of cytochalasin-induced vacuole formation, an ultrastructural study employing Ruthenium red. Tissue Cell 18:685–700

Davis WL, Jones RG, Ciumei J, Knight JP, Goodman DBP (1982) Electron-microscopic and morphometric study of vesiculation in the epithelia cell layer of the toad urinary bladder. Cell Tissue Res 225:619–631

Dawson AB (1948) Variation in the number and size of nuclei in the cells of the kidney tubules of an Australian desert frog, *Cyclorana* (*Chiroleptes*) *alboguttatus* (Günther). Anat Rec 102:393–407

Dawson AB (1951) Functional and degenerate or rudimentary glomeruli in the kidney of two species of Australian frog, *Cyclorana* (*Chiroleptes*) *platycephalus* and *alboguttatus* (Günther). Anat Rec 109:417–429

De Neff SJ, Sever DM (1977) Ontogenetic changes in phototactic behavior of *Ambystoma tigrinum tigrinum* (Amphibia: Urodela). Proc Indiana Acad Sci 86:478–481

De Piceis Polver P, Barni S, Nano R (1981) Ultrastructure of the urinary bladder of *Salamandra salamandra* (L.) (Amphibia Caudata). Monit Zool Ital (NS) 15:123–132

De Piceis Polver P, Fenoglio C, Gerzeli G (1985) Cytochemical study of K^+-dependent p-nitrophenyl-phosphatase activity in the urinary bladder of *Salamandra salamandra*. Acta Histochem (Jena) 76:235–243

Debnam ES, Snart RS (1975) Water transport response of toad bladder to prolactin. Comp Biochem Physiol [A] 52:75–76

Degani G (1981a) The adaptation of *Salamandra salamandra* (L.) from different habitats to terrestrial life. Br J Herpetol 6:169–172

Degani G (1981b) Salinity tolerance and osmoregulation in *Salamandra salamandra* (L.) from different populations. J Comp Physiol [B] 145:133–137

Degani G (1982a) Water balance of salamanders (*Salamandra salamandra* (L.)) from different habitats. Amphibia Reptilia 2:309–314

Degani G (1982b) Temperature tolerance in three populations of salamanders, *Salamandra salamandra* L. Br J Herpetol 6:186–187

Degani G (1982c) Amphibian tadpole interaction in a winter pond. Hydrobiologia 96:3–7

Degani G (1982d) The response to substrate of *Triturus v. vittatus* (Jenyns) (Amphibia, Urodela). Biol Behav 3:215–220

Degani G (1984a) Temperature selection in *Salamandra salamandra* (L.) larvae and juveniles from different habitats. Biol Behav 9:175–183

Degani G (1984b) The selection of hiding places by *Triturus vittatus vittatus* (Jenyns). Biol Behav 9:235–242

Degani G (1985a) Urea tolerance and osmoregulation in *Bufo viridis* and *Rana ridibunda*. Comp Biochem Physiol [A] 82:833–836

Degani G (1985b) Water balance and body fluids of *Salamandra salamandra* (L.) in their natural habitats in summer and winter. Comp Biochem Physiol [A] 82:479–482

Degani G (1986) Growth and behaviour of six species of amphibian larvae in a winter pond in Israel. Hydrobiologia 140:5–10

Degani G (1993) Cannibalism among *Salamandra salamandra* (L.) larvae. Isr J Zool 39:125–129

Degani G (1994) Ecophysiology of *Salamandra salamandra* at the southern limit of its distribution. Mertensiella 4:111–124

Degani G, Meltzer A (1988) Oxygen consumption of a terrestrial toad (*Bufo viridis*) and semi-aquatic frog (*Rana ridibunda*). Comp Biochem Physiol [A] 89:347–349

Degani G, Mendelssohn H (1981a) Seasonal activity of adult and juvenile *Salamandra salamandra* at the southern limit of their distribution. Br J Herpetol 6:79–81

Degani G, Mendelssohn H (1981b) Moisture as a factor influencing selection of hiding places by juvenile *Salamandra salamandra* (L.) from semi-arid habitats. In: Shuval H (ed) Developments in arid zone ecology and environmental quality. Balaban, Rehovot, pp 49–56

Degani G, Mendelssohn H (1982) Seasonal activity of *Salamandra salamandra* (L.) (Amphibia: Urodela: Salamandridae) in the headwaters of the Jordan River. Isr J Zool 31:77–85

Degani G, Mendelssohn H (1983) The habitats, distribution and life history of *Triturus vittatus vittatus* (Jenyns) in the Mount Meron area (Upper Galilee, Israel). Br J Herpetol 6:317–319

Degani G, Nevo E (1986) Osmotic stress and osmoregulation of tadpoles and juveniles in *Pelobates syriacus*. Comp Biochem Physiol [A] 83:365–370

Degani G, Warburg MR (1978) Population structure and seasonal activity of the adult *Salamandra salamandra* (L.) (Amphibia, Urodela, Salamandridae) in Israel. J Herpetol 12:437–444

Degani G, Warburg MR (1980) The response to substrate moisture of juvenile and adult *Salamandra salamandra* (L.) (Amphibia; Urodela). Biol Behav 5:281–290

Degani G, Warburg MR (1984) Changes in the concentrations of ions and urea in both plasma and muscle tissue in a dehydrated hylid anuran. Comp Biochem Physiol [A] 77:357–360

Degani G, Goldenberg S, Warburg MR (1980) Cannibalistic Phenomena in *Salamandra salamandra* larvae in certain water bodies and under experimental conditions. Hydrobiologia 75:123–128

Degani G, Goldenberg S, Warburg MR (1983) Changes in ion, urea concentrations and blood plasma osmolarity of *Pelobates syriacus* juveniles under varying conditions. Comp Biochem Physiol [A] 75:619–623

Degani G, Silanikove N, Shkolnik A (1984) Adaptation of green toad (*Bufo viridis*) to terrestrial life by urea accumulation. Comp Biochem Physiol [A] 77:585–587

Delson J, Whitford WG (1973a) Critical thermal maxima in several life history stages in desert and montane populations of *Ambystoma tigrinum*. Herpetologica 29:352–355

Delson J, Whitford WG (1973b) Adaptation of the tiger salamander, *Ambystoma tigrinum*, to arid habitats. Comp Biochem Physiol [A] 46:631–638

Dent JN (1975) Integumentary effects of prolactin in the lower vertebrates. Am Zool 15:923–935

Dent JN (1988) Hormonal interaction in amphibian metamorphosis. Am Zool 28:297–308

Deyrup IJ (1964) Water balance and kidney. In: Moore J (ed) Physiology of Amphibia. Academic Press, New York, pp 251–328

Diaz NF, Valencia J (1985) Microhabitat utilization by two leptodacylid frogs in the Andes of central Chile. Oecologia 66:353–357

Diaz-Paniagua C (1992) Variability in timing of larval season in an amphibian community in SW Spain. Ecography 15:267–272

Dietz TH, Kirschner LB, Porter D (1967) The roles of sodium transport and anion permeability in generating transepithelial potential differences in larval salamanders. J Exp Biol 46:85–96

Dimmit MA, Ruibal R (1980a) Exploitation of food resources by spadefoot toads (*Scaphiopus*). Copeia 1980:854–862

Dimmit MA, Ruibal R (1980b) Environmental correlates of emergence in spadefoot toads (*Scaphiopus*). J Herpetol 14:21–29

Dineen CF (1955) Food habits of the larval tiger salamander (*Ambystoma tigrinum*). Indiana Acad Sci 65:231–233

Dole JW, Rose BB, Tachiki KH (1981) Western toads (*Bufo boreas*) learn odor of prey insects. Herpetologica 37:63–68

Domm AJ, Janssens PA (1971) Nitrogen metabolism during developoment in the corroboree frog, *Pseudophryne corroboree* Moore. Comp Biochem Physiol [A] 38:163–173

Donnelly MA, Guyer C (1994) Patterns of reproduction and habitat use in an assemblage of neotropical hylid frogs. Oecologia 98:291–302

Douglas ME (1979) Migration and sexual selection in *Ambystoma jeffersonianum*. Can J Zool 57:2303–2310

Douglas ME (1981) A comparative study of topographical orientation in *Ambystoma* (Amphibia: Caudata). Copeia 1981:460–463

Drewes RC, Hillman SS, Putnam RW, Sokol OM (1977) Water, nitrogen and ion balance in the African treefrog *Chiromantis petersi* Boulenger (Anura: Rhacophoridae), with comments on the structure of the integument. J Comp Physiol [B] 116:257–267

Duellman WE (ed) (1979) The South American herpetofauna: its origin, evolution, and dispersal. Monogr Mus Nat Hist Univ Kansas 7

Duellman WE, Trueb L (1986) Biology of amphibians. McGraw-Hill, New York, 670 pp

Duvall D, Norris DO (1980) Stimulation of terrestrial-substrate preferences and locomotor activity in newly transformed tiger salamanders (*Ambystoma tigrinum*) by exogenous or endogenous thyroxine. Anim Behav 28:116–123

Eberth CJ (1869) Untersuchungen zur normalen und pathologischen Anatomie der Froschaut. W Engelmann, Leipzig

Eddy LJ, Allen RF (1979) Prolactin action on short circuit current in the developing tadpole skin: a comparison with ADH. Gen Comp Endocrinol 38:360–364

Elgar MA, Crespi BJ (1992) Ecology and evolution of cannibalism. In: Elgar MA, Crespi BJ (eds) Cannibalism. Oxford University Press, Oxford, pp 1–12

Elkan E (1968) Mucopolysaccharides in the anuran defence against desiccation. J Zool (Lond) 155:19–53

Elkan E (1976) Ground substance: anuran defense against desiccation. In: Lofts B (ed) Physiology of the Amphibia, vol III. Academic Press, New York, pp 101–110

Elkan E, Cooper JE (1980) Skin biology of reptiles and amphibians. Proc R Soc Edinburgh [Biol Sci] 79:115–125

Elliot AB (1968) Natriferic and hydrosmotic effects of neurohypophysial peptides and their analogues in augmenting fluid uptake by *Bufo melanostictus*. J Physiol (Lond) 197:173–182

Espina S, Rojas M (1972) A comparison of the size of the urinary bladder of two South American anurans of different habitat. Comp Biochem Physiol [A] 41:115–119

Espina S, Salibian A, Rojas M (1980) Nitrogen excretion in the South American aquatic leptodactylid *Caudiverbera caudiverbera* (L.). Comp Biochem Physiol [A] 65:487–488

Etheridge K (1990a) Water balance in estivating sirenid salamanders (*Siren lacertina*). Herpetologica 46:400–406

Etheridge K (1990b) The energetics of estivating sirenid salamanders (*Siren lacertina* and *Pseudobranchus striatus*). Herpetologica 46:407–414

Evans DH (1990) An emerging role for a cardiac peptide hormone in fish osmoregulation. Annu Rev Physiol 52:43–60

Ewer RF (1951) Water uptake and moulting in *Bufo regularis* Reuss. J Exp Biol 28:369–373

Ewer RF (1952a) The effect of pituitrin on fluid distribution in *Bufo regularis* Reuss. J Exp Biol 29:173–177

Ewer RF (1952b) The effects of posterior pituitary extracts on water balance in *Bufo carens* and *Xenopus laevis*, together with some general considerations of anuran water economy. J Exp Biol 29:429–439

Fair JW (1970) Comparative rates of rehydration from soil in two species of toads *Bufo boreas* and *Bufo punctatus*. Comp Biochem Physiol 34:281–287

Fanelli GM, Goldstein L (1964) Ammonium excretion in the neotenous newt *Necturus maculosus* (Rafinesque). Comp Biochem Physiol 13:193–204

Farrell MP, MacMahon JA (1969) An eco-physiological study of water economy in eight species of tree frogs (Hylidae). Herpetologica 25:279–294

Fasola M (1993) Resource partitioning by three species of newts during their aquatic phase. Ecography 16:73–81

Faszweski EE, Kaltenbach JC (1995) Histology and lectin-binding patterns in the skin of the terrestrial horned frog *Ceratophrys ornata*. Cell Tissue Res 281:169–177

Feder ME, Burggren WW (1985) Cutaneous gas exchange in vertebrates: design, patterns, control and implications Biol Rev 60:1–45

Feder ME, Burggren WW (1992) Environmental physiology of the amphibians. University of Chicago Press, Chicago, 646 pp

Ferguson DE (1967) Sun-compass orientation in anurans. In: Storm RM (ed) Animal orientation and navigation. Oregon State University Press, Corvallis, pp 21–33

Ferguson DE, Landreth HF (1966) Celestial orientation of Fowler's toad *Bufo fowleri*. Behaviour 26:107–123

Fernandez PJ (1988) Purine and pteridine pigments of light- and dark-colored Arizona tiger salamanders. J Exp Zool 245:121–129

Fernandez PJ, Collins JP (1988) Effect of environment and ontogeny on color pattern variation in Arizona tiger salamanders (*Ambystoma tigrinum nebulosum* Hollowell). Copeia 1988:928–938

Fitzgerald GJ, Bider JR (1974) Evidence for a relationship between geotaxis and seasonal movements in the toad *Bufo americanus*. Oecologia 17:277–280

Flanigan JE, Withers PC, Guppy M (1991) In vitro metabolic depression of tissues from the aestivating frog *Neobatrachus pelobatoides*. J Exp Biol 161:273–283

Floyd RB (1983) Ontogenetic change in the temperature tolerance of larval *Bufo marinus* (Anura: Bufonidae). Comp Biochem Physiol [A] 75:267–271

Floyd RB (1984) Variation in temperature preference with stage of development of *Bufo marinus* larvae. J Herpetol 18:153–158

Floyd RB (1985) Effects of photoperiod and starvation on the temperature tolerance of larvae of the giant toad, *Bufo marinus*. Copeia 1985:625–631

Forge P, Barbault R (1978) Observations sur le régime alimentaire de deux amphibiens sympatriques du Sahel Senegalais: *Bufo pentoni* et *Tomopterna delalandi*. Bull IFAN 40:674–684

Formas JR (1979) La herpetofauna de los bosques temperados de Sudamerica. Monogr Mus Nat Hist Kansas 7:341–369

Fox H (1963) The amphibian pronephros. Q Rev Biol 38:1–25

Francillon H, Barbault R, Castanet J, de Ricqles A (1984) Étude complemmentaire sur la biologie de l'amphibien deserticole *Bufo pentoni*: Données de squelettochronologie et d'écodemographie. Rev Ecol (Terre Vie) 39:209–224

Frazier LW (1983) The effect of catecholamines on H^+ and NH_4^+ excretion in toad urinary bladder. Comp Biochem Physiol [C] 75:321–326

Frazier LW, Vanatta JC (1980) Evidence that the frog skin excretes ammonia. Comp Biochem Physiol [A] 66:525–527

Frazier LW, Vanatta JC (1981) Excretion of K^+ by frog skin with rate varying with K^+ load. Comp Biochem Physiol [A] 69:157–160

Frost DR (1985) Amphibian species of the world. Allen Ross, Lawrence, Kansas, 732 pp

Funkhouser A (1977) Plasma osmolarity of *Rana catesbeiana* and *Scaphiopus hammondi* tadpoles. Herpetologica 33:272–274

Galey WR, Wood SC, Mancha VM (1987) Effects of metamorphosis on water permeability of skin in the salamander, *Ambystoma tigrinum*. Comp Biochem Physiol [A] 86:429–432

Gallardo JM (1979) Composicion, distribution y origen de la herpetofauna Chaquena. Monogr Mus Nat Hist Univ Kansas 7:299–307

Garcia-Romeu F, Salibian A (1968) Sodium uptake and ammonia excretion through the in vivo skin of the South American frog *Leptodactylus ocellatus* (L., 1758). Life Sci 7:465–470

Garland HO, Henderson IW, Brown JA (1975) Micropuncture study of the renal resposes of the urodele amphibian *Necturus maculosus* to injection of arginine vasotocin and an anti aldosterone compound. J Exp Biol 63:249–264

Gascon C (1992) The effects of reproductive phenology on larval performance traits in a three-species assemblage of central Amazonian tadpoles. Oikos 65:307–313

Gasser KW, Miller BT (1986) Osmoregulation of larval blotched tiger salamanders, *Ambystoma tigrinum melanostictum*, in saline environments. Physiol Zool 59:643–648

Gealekman O (1996) Changes in the presence of atrial natriuretic peptides (ANP) in the heart of *Pelobates syriacus* during ontogenesis and under hyperosmotic conditions. Isr J Zool (in press)

Gehlbach FR, Gordon R, Jordan JB (1973) Aestivation of the salamander, *Siren intermedia*. Am Midl Nat 89:455–463

Gehlbach FR, Kimmel JR, Weems WA (1969) Aggregations and body water relations in tiger salamanders (*Ambystoma tigrinum*) from the Grand Canyon rims, Arizona. Physiol Zool 42:173–182

Geise W, Linsenmair KE (1986) Adaptations of the reed frog *Hyperolius viridiflavus* (Amphibia, Anura, Hyperoliidae) to its arid environment. II. Some aspects of the water economy of *Hyperolius viridiflavus nitidulus* under wet and dry season conditions. Oecologia 68:542–548

Geise W, Linsenmair KE (1988) Adaptations of the reed frog *Hyperlius viridiflavus* (Amphibia, Anura, Hyperoliidae) to its arid environment. IV. Ecological significance of water economy with comments on thermoregulation and energy allocation. Oecologia 77:327–338

Gern WA, Norris DO (1979) Plasma melatonin in the neotenic tiger salamander (*Ambystoma tigrinum*): effects of photoperiod and pinealectomy. Gen Comp Endocrinol 38:393–398

Gern WA, Norris DO, Duvall D (1983) The effect of light and temperature on plasma melatonin in neotenic tiger salamander (*Ambystoma tigrinum*). J Herpetol 17:228–234

Gibbons JW, Bennett DH (1974) Determination of anuran terrestrial activity patterns by a drift fence method. Copeia 1974:236–243

Giorgio M, Giacoma C, Vellano C, Mazzi V (1982) Prolactin and sexual behaviour in the crested newt (*Triturus cristatus carnifex* Laur.). Gen Comp Endocrinol 47:139–147

Glauert L (1945) Some Western Australian frogs. Aust Mus Mag 1945:379–382

Goetz KL (1988) Physiology and pathophysiology of atrial peptides. Am J Physiol 254:E1–E15

Goin CJ (1960) Amphibians, pioneers of terrestrial breeding habits. Smithson Rep 4404:427–445

Goldenberg S, Warburg MR (1976) Changes in the response to oxytocin followed throughout ontogenesis in two anuran species. Comp Biochem Physiol [C] 53:105–113

Goldenberg S, Warburg MR (1977a) Osmoregulatory effect of prolactin during ontogenesis in two anurans. Comp Biochem Physiol [A] 56:137–143

Goldenberg S, Warburg MR (1977b) Changes in the effect of vasotocin on water balance of *Rana ridibunda* during ontogenesis. Comp Biochem Physiol [A] 57:451–455

Goldenberg S, Warburg MR (1983) Water balance of five amphibian species at different stages and phases, as affected by hypophysial hormones. Comp Biochem Physiol [A] 75:447–455

Goldman JM, MacHadley E (1969) The beta adrenergic receptor and cyclic 3',5'-adenosine monophosphate: possible roles in the regulation of melanophore responses of the spadefoot toad, *Scaphiopus couchi*. Gen Comp Endocrinol 13:151–163

Gona O (1982) Uptake of I 125-labelled prolactin by bullfrog kidney tubules: an autoradiographic study. J Endocrinol 93:193–198

Goniakowska-Witalinska L (1978) Ultrastructural and morphometric study of the lung of the European salamander, *Salamadra salamandra* L. Cell Tissue Res 191:343–356

Goniakowska-Witalinska L (1980a) Scanning and transmission electron microscopic study of the lung of the newt, *Triturus alpestris* Laur. Cell Tissue Res 205:133–145

Goniakowska-Witalinska L (1980b) Ultrastructural and morphometric changes in the lung of the newt *Triturus cristatus carnifex* Laur. during ontogeny. J Anat 130:571–583

Goniakowska-Witalinska L (1980c) A peculiar mode of formation of the surface lining layer in the lungs of *Salsmandra salamandra*. Tissue Cell 12:539–546

Goniakowska-Witalinska L (1980d) Endocrine-like cells in the lungs of the newt, *Triturus alpestris* Laur. Cell Tissue Res 210:521–524

Goniakowska-Witalinska L (1980e) A new type of cell in the pulmonary epithlium of the newt *Triturus alpestris* Laur. Cell Tissue Res 210:479–484

Goniakowska-Witalinska L (1982) Development of the larval lung of *Salamandra salamandra* L. Anat Embryol (Berl) 164:113–137

Goniakowska-Witalinska L, Cutz E (1990) Ultrastructure of neuroendocrine cells in the lungs of the anuran species. J Morphol 203:1–9

Goniakowska-Witalinska L, Lauweryns JM, Zaccone G, Fasulo S, Tagliafierro G (1992) Ultrastructure and immunocytochemistry of the neuroepithelial bodies in the lung of the tiger salamander, *Ambystoma tigrinum* (Urodela, Amphibia). Anat Rec 234:419–431

Gordon MS (1962) Osmotic regulation in the green toad (*Bufo viridis*). J Exp Biol 39:261–270

Gosner KL, Black IH (1955) The effects of temperature and moisture on the reproductive cycle of *Scaphiopus h. holbrooki*. Am Midl Nat 54:192–203

Grafe TU, Schmuch R, Linsenmair KE (1992) Reproductive energetics of the African reed frogs, *Hyperolius viridiflafus and Hyperolius marmoratus*. Physiol Zool 65:153–171

Grant WC (1961) Special aspects of the metamorphic process: second metamorphosis. Am Zool 1:163–171

Grassi Milano E, Accordi F (1983) Comparative morphology of the adrenal gland of anuran Amphibia. J Anat 136:165–174

Grassi Milano E, Accordi F (1986) Evolutionary trends in adrenal gland of anurans and urodeles. J Morphol 189:249–259

Grassi Milano E, Accordi F (1991) Morphological evolution of the adrenal gland of amphibians. In: Ghiara G et al. (eds) Symposium on the evolution of terrestrial vertebrates. Selected symposia and monographs U Z I, vol 4. Mucchi, Modena, pp 487–495

Gray JE (1845) Description of some new Australian animals. In: Journals of Expeditions of Discovery into Central Australia and Overland, vol I. Eyre EJ, Boone T & W Publ, London

Greven H (1976) Notizen zum Geburtsvorgang beim Feuersalamander, *Salamandra salamandra* (L.). Salamandra 12:87–93

Greven H (1980) Ultrastructural investigations of the epidermis and the gill epithelium in the intrauterine larvae of *Salamandra salamandra* (L.) (Amphibia, Urodela). Z Mikrosk Anat Forsch 94:196–208

Greven H, Guex GD (1994) Structural and physiological aspects of viviparity in *Salamandra salamandra*. Mertensiella 4:139–160

Griffiths RA (1984) A comparative study of phototaxis and the response to substrate moisture in newts and salamanders. Br J Herpetol 6:375–378

Grodzinski Z (1975) The morphology of yolk platelets in the developing egg of *Salamandra salamandra* L. Acta Biol Cracov Ser Zool 18:131–145

Grosse WR, Bauch S (1988) Zum Wachstum der Kaulquappen und zur Lungenentwicklung des Laubfrosches (*Hyla a aroborea* [L.]). Zool Abh Staatl Mus Tiere (Dresd) 44:11–18

Grosse WR, Linnenbach M (1989) Feinbau der Hautdrüsen und Entwicklung des Melanophoren-Musters beim Laubfrosch, *Hyla arborea* L. (Amphibia, Anura). Zool Anz 223:211–219

Grubb JC (1970) Orientation in post-reproductive Mexican toads, *Bufo valliceps*. Copeia 1970:674–680

Grubb JC (1973a) Olfactory orientation in breeding Mexican toads, *Bufo valliceps*. Copeia 1973:490–497

Grubb JC (1973b) Orientation in newly metamorphosed Mexican toads, *Bufo valliceps*. Herpetologica 29:95–100

Grubb JC (1976) Maze orientation by Mexican toads, *Bufo valliceps* (Amphibia, Anura, Bufonidae), using olfactory and configurational cues. J Herpetol 10:97–104

Guardabassi A (1960) The utilization of the calcareous deposits in the endolymphatic sacs of *Bufo bufo bufo* in the mineralization of the skeleton. Investigations by means of ^{45}Ca. Z Zellforsch Mikrosk Anat 51:278–282

Guix JC (1993) Habitat y alimentacion de *Bufo paracnemis* en una region semiarida del nordeste de Brasil, durante el periodo de reproduccion. Rev Esp Herpetol 7:65–73

Hailman JP (1984) Bimodal nocturnal activity of the western toad (*Bufo boreas*) in relation to ambient illumination. Copeia 1984:283–290

Hailman JP, Jaeger RG (1974) Phototactic responses to spectrally dominant stimuli and use of color vision by adult anuran amphibians: a comparative survey. Anim Behav 22:757–795

Han DS, Park YS, Hong SK (1978) Seasonal variation in energy metabolism and neurohypophyseal hormone action on water and sodium transport in frogs (*Rana temporaria*) and toads (*Bufo marinus*). Comp Biochem Physiol [A] 61:665–669

Hanke W (1985) A comparison of endocrine function in osmotic and ionic adaptation in amphibians and teleost fishes. In: Follett BK, Ishii S, Chandola A (eds) The endocrine system and the environment. Japanese Scientific Societies Press, Tokyo, pp 33–43

Hanke W, Kloas W (1994) Hormonal regulation of osmomineral content in Amphibia. Zool Sci 11:5–14

Hardy LM, Raymond LR (1980) The breeding migration of the mole salamander, *Ambystoma talpoideum*, in Louisiana. J Herpetol 14:327–335

Harrison L (1922) On the breeding habits of some Australian frogs. Aust Zool 3:17–34

Heath AG (1976) Respiratory responses to hypoxia by *Ambystoma tigrinum* larvae, paedomorphs, and metamorphosed adults. Comp Biochem Physiol [A] 55:45–49

Heatwole H (1984) Adaptations of amphibians to aridity. In: Cogger HG, Cameron EE (eds) Arid Australia Australian Museum, Sydney, pp 177–222

Heatwole H, Cameron E, Webb GJW (1971) Studies on anuran water balance. II. Desiccation in the Australian frog *Notaden bennetti*. Herpetologica 27:365–378

Heatwole H, Mercado N, Ortiz E (1965) Comparison of critical thermal maxima of two species of Puerto Rican frogs of the genus *Eleutherodactylus*. Physiol Zool 38:1–8

Heatwole H, Newby RC (1972) Interaction of internal rhythm and loss of body water in influencing activity levels of amphibians. Herpetologica 28:156–162

Heatwole H, Blasini de Austin S, Herrero R (1968) Heat tolerances of two tadpoles of two species of tropical anurans. Comp Biochem Physiol 27:807–815

Heatwole H, Torres F, de Austin SB, Heatwole A (1969) Studies on anuran water balance. I. Dynamics of evaporative water loss by the Coqui, *Eleutherdactylus portoricensis*. Comp Biochem Physiol 28:245–269

Heinen JT (1993) Aggregations of newly metamorphosed *Bufo americanus*: tests of two hypotheses. Can J Zool 71:334–338

Heller H (1965) Osmoregulation in Amphibia. Arch Anat Microsc Morphol Exp 54:471–490

Heller H (1970) The neuro-endocrine control of water metabolism. Mem Soc Endocrinol 18:447–463

Heller H, Bentley PJ (1965) Phylogenetic distribution of the effects of neurohypophysial hormones on water and sodium metabolism. Gen Comp Endocrinol 5:96–108

Henderson IW, Kime DE (1987) The adrenal cortical steroids. In: Pang PKT, Schreibman MP (eds) Vertebrate endocrinology: fundamentals and biomedical implications, vol 2. Academic Press, New York, pp 121–142

Heney HW, Stiffler DF (1983) The effect of aldosterone on sodium and potassium metabolism in larval *Ambystoma tigrinum*. Gen Comp Endocrinol 49:122–127

Herter K (1941) Die Physiologic der Amphibien. In: Kükenthal W (ed) Handbuch der Zoologie, vol VI. Walter de Gruyter, Berlin

Hetherington TE, Lombard RE (1983) Mechanisms of underwater hearing in larval and adult tiger salamanders *Ambystoma tigrinum*. Comp Biochem Physiol [A] 74:555–559

Heusser H (1960) Über die Beziehungen der Erdkröte, *Bufo bufo* L., zu ihrem Laichplatz II. Behaviour 16:94–109

Heyer RW (1969) The adaptive ecology of the species groups of the genus *Leptodactylus* (Amphibia, Leptodactylidae). Evolution 23:421–428

Hickman CP, Trump BF (1969) The kidney. In: Randall DJ, Hoar WS (eds) Fish Physiology, vol 1. Academic Press, New York, pp 91–230

Higginbotham AC (1939) Studies on the amphibian activity. I. Preliminary report on the rhythmic activity of *Bufo americanus* Holbrook and *Bufo fowleri* Hinckley. Ecology 20:58–70

Hillman SS (1976) Cardiovascular correlates of maximal oxygen consumption rates in anuran amphibians. J Comp Physiol 109:199–207

Hillman SS (1980) Physiological correlates of differential dehydration tolerance in anuran amphibians. Copeia 1980:125–129

Hillman SS (1987) Dehydrational effects on cardiovascular and metabolic capacity in two amphibians. Physiol Zool 60:608–613

Hillman SS, Zygmunt A, Baustian M (1987) Transcapillary fluid forces during dehydration in two amphibians. Physiol Zool 60:339–345

Hillyard SD (1975) The role of antidiuretic hormones in the water economy of the spadefoot toad, *Scaphiopus couchi*. Physiol Zool 48:242–251

Hillyard SD (1976a) Variation in the effects of antidiuretic hormone on isolated skin of the toad, *Scaphiopus couchi*. J Exp Zool 195:199–206

Hillyard SD (1976b) The movement of soil water across the isolated amphibian skin. Copeia 1976:314–320

Hillyard SD (1979) The effect of isoproterenol on the anuran water balance response. Comp Biochem Physiol [C] 62:93–95

Himstedt W (1971) Die Tagesperiodik von Salamandriden. Oecologia 8:194–208

Himstedt W (1973) Die spektrale Empfindlichkeit von Urodelen in Abhängigkeit von Metamorphose, Jahreszeit und Lebensraum. Zool JB Physiol 77:246–274

Himstedt W (1982) Evolutionary aspects of color vision in amphibians. In: Mossakowski D, Roth G (eds) Environmental adaptation and evolution. Fischer, Stuttgart, pp 67–85

Himstedt W (1994) Sensory systems and orientation in *Salamandra salamandra*. Mertensiella 4:225–239

Himstedt W, Plasa L (1979) Home-site orientation by visual cues in salamanders. Naturwissenschaften 66:372

Himstedt W, Freidank U, Singer E (1976) Die Veränderung eines Auslösemechanismus im Beutefangverhalten während der Entwicklung von *Salamandra salamandra* (L.). Z Tierpsychol 41:235–243

Himstedt W, Tempel P, Weiler J (1978) Responses of salamanders to stationary visual patterns. J Comp Physiol 124:49–52

Hirano T (1977) Prolactin and hydromineral metabolism in the vertebrates. Gumna Symp Endocrinol 14:45–59

Hirano T, Ogasawara T, Bolton JP, Collie NL, Hasegawa S, Iwata M (1987) Osmoregulatory role of prolactin in lower vertebrates. In: Kirsch, Lalou (eds) Comparative physiology of environmental adaptations, vol 1. Karger, Basel, pp 112–124

Hirohama T, Uemura H, Nakamura S, Naruse M, Aoto T (1989) Ultrastructure and atrial natriuretic peptide (ANP)-like immunoreactivity of cardiocytes in the larval, metamorphosing and adult specimens of the Japanese toad *Bufo japonicus formosus*. Dev Growth Diff 31:113–121

Hödl W (1975) Die Entwicklung der spektralen Empfindlichkeit der Netzhaut von *Bombina*, *Hyla*, *Pelobates* und *Rana*. Zool Jahrb Physiol 79:173–203

Hoff Seckendorff K von, Hillyard SD (1991) Angiotensin II stimulates cutaneous drinking in the toad *Bufo punctatus*. Physiol Zool 64:1165–1172

Hoff Seckendorff K von, Hillyard SD (1993a) Inhibition of cutaneous water absorption in dehydrated toads by saralasin is associated with changes in barometric pressure. Physiol Zool 66:89–98

Hoff Seckendorff K von, Hillyard SD (1993b) Toad taste sodium with their skin: sensory function in a transporting epithlium. J Exp Biol 183:347–351

Hoffman CW, Dent JN (1978) The morphology of the mucous gland and its responses to prolactin in the skin of the redspotted newt. J Morphol 157:79–98

Hoffman J, Katz U (1989) The ecological significance of burrowing behaviour in the toad (*Bufo viridis*). Oecologia 81:510–513

Hoffman J, Katz U (1991) Tissue osmolytes and body fluid compartments in the toad *Bufo viridis* under simulated terrestrial conditions. J Comp Physiol [B] 161:433–439

Hoffman J, Katz U (1994) Urea production and accumulation in the green toad, *Bufo viridis*. J Zool (Lond) 233:591–603

Hoffman J, Katz U, Eylath U (1990) Urea accumulation in response to water restriction in burrowing toads (*Bufo viridis*). Comp Biochem Physiol [A] 97:423–426

Holloway WR, Dapson RW (1971) Histochemistry of integumentary secretions of the narrowmouthed toad, *Gastrophryne carolinensis*. Copeia 1971:351–353

Holomuzki JR (1986) Predator avoidance and diel patterns of microhabitat use by larval tiger salamanders. Ecology 67:737–748

Honjo J (1939) Farbensinn der Feuersalamanderlarve. Men Coll Sci Kyoto Imp Univ 15:207

Hota AK, Dash MC (1981) Growth and metamorphosis of *Rana tigrina* larvae: effects of food level and larval density. Oikos 37:349–352

Hourdry J (1993) Passage to the terrestrial life in amphibians. II. Endocrine determinism. Zool Sci 10:887–902

Hovingh P, Benton B, Bornholdt D (1985) Aquatic parameters and life history observations of the Great Basin spadefoot toad in Utah. Great Basin Nat 45:22–30

Howard JH, Wallace RL (1984) Life history characteristics of populations of the long-toed salamander (*Ambystoma macrodactylum*) from different altitudes. Am Midl Nat 113:361–373

Hughes GM (1966) Species variation in gas exchange. Proc R Soc Med 59:494–500

Hughes GM (1967) Evolution between air and water. In: de Reuck AVS, Porter R (eds) Development of the lung. Little Brown, Boston, pp 64–80

Huja BS, Hong SK (1976) Characteristics of vasopressin action on Na transport across the isolated toad skin. Comp Biochem Physiol [A] 53:187–191

Hunsaker D, Johnson C (1959) Internal pigmentation and ultraviolet transmission of the integument in amphibians and reptiles. Copeia 1959:311–315

Husting EL (1965) Survival and breeding structure in a population of *Ambystoma maculatum*. Copeia 1965:352–362

Hutchison VH (1989) Annual cycle of thermal tolerance in the salamander, *Necturus maculosus*. J Herpetol 23:73–76

Hutchison VH, Dupre RK (1992) Thermoregulation. In: Feder ME, Burggren WW (eds) Environmental physiology of the amphibians. University of Chicago Press, Chicago, pp 206–249

Hutchison VH, Kohl MA (1971) The effect of photoperiod on daily rhythms of oxygen consumption in the tropical toad, *Bufo marinus*. Z Vergl Physiol 75:367–382

Hutchison VH, Whitford WG, Kohl M (1968) Relation of body size and surface area to gas exchange in anurans. Physiol Zool 41:65–85

Hutchison VH, Turney LD, Gratz RK (1977) Aerobic and anaerobic metabolism during activity in the salamander *Ambystoma tigrinum*. Physiol Zool 50:189–202

Iela L, Rastogi RK, Delrio G, Bagnara JT (1986) Reproduction in the Mexican leaf frog, *Pachymedusa danicolor*. III. The female. Gen Comp Endocrinol 63:381–392

Imamura A, Takeda H, Sasaki N (1965) The accumulation of sodium and calcium in a specific layer of frog skin. J Cell Comp Physiol 66:221–226

Ingle D, McKinley D (1978) Effects of stimulus configuration on elicited prey catching by the marine toad (*Bufo marinus*). Anim Behav 26:885–891

Ireland MP (1973) Effects of arginine vasotocin on sodium and potassium metabolism in *Xenopus laevis* after skin gland stimulation and sympathetic blockade. Comp Biochem Physiol [A] 44:487–493

Ireland MP, Simons IM (1977) Adaptation of the axolotl (*Ambystoma mexicanum*) to a hyperosmotic medium. Comp Biochem Physiol [A] 56:415–417

Ishii S, Kubokawa K, Kikuchi M, Nishio H (1995) Orientation of the toad, *Bufo japonicus*, towards the breeding pond. Zool Sci 12:475–484

Itoh M, Ishii S (1990) The relationship between amplexus behavior and plasma gonadotropin and sex steroid levels in male toads, *Bufo japonicus*. Zool Sci 7:1139

Jackson BA, Henderson IW (1976) Arginine-vasotocin actions on renal tubular and glomerular function in *Bufo marinus*. J Endocrinol 71:78P

Jaeger RG, Hailman JP (1973) Effects of intensity on the phototactic responses of adult anuran amphibians: a comparative survey. Z Tierpsychol 33:352–407

Jaffe RC (1981) Plasma concentration of corticosterone during *Rana catesbeiana* tadpole metamorphosis. Gen Comp Endocrinol 44:314–318

Jameson DL (1955) The population dynamics of the cliff frog, *Syrrhophus marnocki*. Am Midl Nat 54:342–381

Jameson DL (1966) Rate of weight loss of tree frogs at various temperatures and humidities. Ecology 47:605–613

Jarial MS (1989) Fine structure of the epidermal Leydig cells in the axolotl *Ambystoma mexicanum* in relation to their function. J Anat 167:95–102

Jasinski A, Gorbman A (1967) Hypothalamic neurosecretion in the spadefoot toad, *Scaphiopus hammondi*, under different environmental conditions. Copeia 1967:271–279

Jennions MD, Passmore NI (1993) Sperm competition in frogs: testis size and a "sterile male" experiment on *Chiromantis xerampelina* (Rhacophoridae). Biol J Linn Soc 50:211–220

Jennions MD, Backwell PRY, Passmore NI (1992) Breeding behaviour of the African frog, *Chiromantis xeramplina*: multiple spawning and polyandry. Anim Behav 44:1091–1100

Johnson CR (1969a) Water absorption response of some Australian anurans. Herpetologica 25:171–172

Johnson CR (1969b) Aggregation as a means of water conservation in juvenile *Limnodynastes* from Australia. Herpetologica 25:275–276

Johnson CR (1970) Observations on body temperatures, critical thermal maxima and tolerance to water loss in the Australian hylid, *Hyla caerulea* (White). Proc R Soc Queensl 82:47–50

Johnson CR (1971a) Thermal relations in some southern and eastern Australian anurans. Proc R Soc Queensl 82:87–94

Johnson CR (1971b) Daily variation in thermal tolerance of *Litoria caerulea* (Anura: Hylida). Comp Biochem Physiol [A] 40:1109–1111

Johnson CR (1972a) Thermal relations and daily variation in the thermal tolerance in *Bufo marinus*. J Herpetol 6:35–38

Johnson CR (1972b) Diel variation in the thermal tolerance of *Litoria gracileata* (Anura: Hylidae). Comp Biochem Physiol [A] 41:727–730

Jolivet-Jaudet G, Ishii S (1985) Annual changes in interrenal function in the Japanese toad, *Bufo japonicus*. In: Follet B Ishii K, Chandola A (eds) The endocrine system and the environment. Springer, Berlin Heidelberg New York, pp 45–53

Jolivet-Jaudet G, Leloup-Hatey J (1984) Variations in aldosterone and corticosterone plasma levels during metamorphosis in *Xenopus laevis* tadpoles. Gen Comp Endocrinol 56:59–65

Jolivet-Jaudet G, Inoue M, Takada K, Ishii S (1984) Circannual changes in plasma aldosterone levels in *Bufo japonicus formosus*. Gen Comp Endocrinol 53:163–167

Joly J (1959) Donées sur l'écologie de la salamandre tachetée: *Salamandra salamandra taeniata* Düringen (1897). Bull Soc Zool Fr 84:208–215

Joly J (1966) Sur l'éthologie sexuelle de *Salamandra salamandra* (L.). Z Tierpsychol 23:8–27

Joly J (1986) La reproduction de la salamandre terrestre (*Salamandra salamandra* L.). In: Grasse PP, Delsol M (eds) Traite de zoologie, Tome 14. Masson, Paris

Joly J, Picheral B (1972) Ultrastructure, histochimie et physiology du follicule pré-ovulatoire et du corps jaune de l'urodèle ovo-vivipaire *Salamandra salamandra*. Gen Comp Endocrinol 18:235–259

Joly J, Chesnel F, Boujard D (1994) Physiological and biological reproductive strategies in the genus *Salamandra*. Mertensiella 4:139–160

Jones JM, Wentzell LA, Toews DP (1992) Posterior lymph heart pressure and rate and lymph flow in the toad *Bufo marinus* in response to hydrated and dehydrated conditions. J Exp Biol 169:207–220

Jones RM (1978) Rapid pelvic water uptake in *Scaphiopus couchi* toadlets. Physiol Zool 51:51–55

Jones RM (1980a) Nitrogen excretion by *Scaphiopus* tadpoles in ephemeral ponds. Physiol Zool 53:26–31

Jones RM (1980b) Metabolic consequences of accelerated urea synthesis during seasonal dormancy of spadefoot toads, *Scaphiopus couchi* and *Scaphiopus multiplicatus*. J Exp Zool 212:255–267

Jones RM (1982) Urea synthesis and osmotic stress in the terrestrial anurans *Bufo woodhousei* and *Hyla cadaverina*. Comp Biochem Physiol [A] 71:293–297

Jones RM, Hillman SS (1978) Salinity adaptation in the salamander *Batrachoseps*. J Exp Biol 76:1–10

Jørgensen CB (1950) The amphibian water economy with special regard to the effect of neurohypophyseal extracts. Acta Physiol Scand 22 Suppl 78:1–79

Jørgensen CB (1984) Ovarian functional patterns in Baltic and Mediterranean populations of a temperate zone anuran, the toad *Bufo viridis*. Oikos 43:309–321

Jørgensen CB (1993a) Role of pars nervosa of the hypophysis in amphibian water economy: a reassessment. Comp Biochem Physiol [A] 104:1–21

Jørgensen CB (1993b) Effects of feeding on water balance and cutaneous drinking in the toad (*Bufo bufo* L.). Comp Biochem Physiol [A] 106:793–798

Jørgensen CB, Larsen LO (1964) Further observations on molting and its hormonal control in *Bufo bufo* (L.). Gen Comp Endocrinol 4:389–400

Jørgensen CB, Levi H, Ussing HH (1947) On the influence of the neurohypophyseal principles on the sodium metabolism in the axolotl (*Amblystoma mexicanum*). Acta Physiol Scand 12:350–370

Jørgensen CB, Larsen LO, Rosenkilde P (1965) Hormonal dependency of molting in amphibians: effect of radiothyroidectomy in the toad *Bufo bufo* (L.). Gen Comp Endocrinol 5:248–251

Jørgensen CB, Larsen LO, Lofts B (1979) Annual cycle of fat bodies and gonads in the toad *Bufo bufo bufo*. K Dan Vidensk Selsk Biol Skr 22(5):1–37

Jørgensen CB, Voûte CL (1979) A possible role of thyrotropin releasing hormone in the seasonal adaptation of salt transport in the frog. Gen Comp Endocrinol 37:482–486

Jungreis AM (1976) Partition of excretory nitrogen in Amphibia. Comp Biochem Physiol [A] 53:133–141

Jurgens JD (1979) The Anura of the Etosha national park. Madoqua 2:185–208

Justus JT, Sandomir M, Urquhart T, Ewan BO (1977) Developmental rates of two species of toads from the desert Southwest. Copeia 1977:592–594

Kaplan RH (1980a) Ontogenetic energetics in *Ambystoma*. Physiol Zool 53:43–56

Kaplan RH (1980b) The implications of ovum size variability for offspring fitness and clutch size within several populations of salamanders (*Ambystoma*). Evolution 34:51–64

Kaplan RH, Salthe SN (1979) The allometry of reproduction: an empirical view in salamanders. Am Nat 113:671–689

Karlstrom EL (1962) The toad genus *Bufo* in the Sierra Nevada of California. Univ Calif Publ Zool 62:1–104

Kasperczyk M (1971) Comparative studies on colour sense in Amphibia (*Rana temporaria* L., *Salamandra salamandra* L. and *Triturus cristatus* Laur.). Folia Biol (Praha) 19:241–288

Kästle W (1986) Rival combats in *Salamandra salamandra*. In: Roček Z (ed) Studies in herpetology. Charles University Prague, pp 525–558

Katschenko N (1882) Über die Krappfärbung der Froschgewebe. Arch Mikr Anat Entw Mech 21:357–386

Katz U (1973) Studies on adaptation of the toad *Bufo viridis* to high salinities: oxygen consumption, plasma concentration and water content of the tissues. J Exp Biol 58:785–796

Katz U (1975) Salt-induced changes in sodium transport across the skin of the euryhaline toad, *Bufo viridis*. J Physiol (Lond) 247:537–550

Katz U (1980) The effect of dehydration on the in vivo acid-base status of the blood in the toad *Bufo viridis*. J Exp Biol 88:403–405

Katz U (1986) The role of amphibian epidermis in osmoregulation and its adaptive response to changing environment. In: Bereiter-Hahn J, Matoltsy AG, Richards KS (eds) Biology of the integument, vol II. The vertebrates. Springer, Berlin Heidelberg New York, pp 472–498

Katz U (1989) Strategies of adaptation to osmotic stress in anuran Amphibia under salt and burrowing conditions. Comp Biochem Physiol [A] 93:499-503

Katz U, Gabbay S (1988) Mitochondria-rich cells and carbonic anhydrase content of toad skin epithelium. Cell Tissue Res 251:425–431

Katz U, Graham R (1980) Water relations in the toad (*Bufo viridis*) and a comparison with the frog (*Rana ridibunda*). Comp Biochem Physiol [A] 67:245–251

Katz U, Hoffman J (1990) Changing plasma osmolality – a strategy of adaptation of anuran Amphibia to water scarcity under burrowing conditions. Fortschr Zool 38:351–356

Katz U, Degani G, Gabbay S (1984) Acclimation of the euryhaline toad *Bufo viridis* to hyperosmotic solution (NaCl, urea and mannitol). J Exp Biol 108:403–409

Katz U, Garcia Romeu F, Masoni A, Isaia J (1981) Active transport of urea across the skin of the euryhaline toad, *Bufo viridis*. Pflügers Arch 390:299–300

Katz U, Pagi D, Hayat S, Degani G (1986) Plasma osmolality, urine composition and tissue water content of the toad *Bufo viridis* Laur, in nature and under controlled laboratory conditions. Comp Biochem Physiol [A] 85:703–713

Kaul R, Shoemaker VH (1989) Control of thermoregulatory evaporation in the waterproof treefrog *Chiromantis xerampelina*. J Comp Biochem [B] 158:643–649

Kawaguti S (1969) Electron microscopy on guanophores in the white belly skin of the tree frog. Annot Zool Jpn 42:8–12

Kawamata S (1990a) Fine structural changes in the endolymphatic sac induced by calcium loading in the tree frog, *Hyla arborea japonica*. Arch Histol Cytol 53:397–404

Kawamata S (1990b) Localization of pyroantimonate-precipitable calcium in the endolymphatic sac of the tree frog, *Hyla arborea japonica*. Arch Histol Cytol 53:405–411

Kawamata S, Takaya K, Yoshida T (1987) Light- and electron-microscopic study of the endolymphatic sac of the tree frog, *Hyla arborea japonica*. Cell Tissue Res 249:57–62

Keen WH, Schroeder EE (1975) Temperature selection and tolerance in three species of *Ambystoma* larvae. Copeia 1975:523–530

Kehr AI, Adema EO (1990) Crecimiento corporal y analisis estadistico de la frecuencia por clases de edades de los estadios larvales de *Bufo arenarum* en condiciones naturales. Neotropica 36:67–81

Kent W, McClanahan LL (1980) The effects of arginine vasotocin and various microtubular poisons on water transfer and sodium transport across the pelvic skin of the toad *Bufo boreas* in vitro. Gen Comp Endocrinol 40:161–167

Kerstetter TH, Kirschner LB (1971) The role of hypothalamo-neurohypophysial system in maintaining hydromineral balance in larval salamanders (*Ambystoma tigrinum*). Comp Biochem Physiol [A] 40:373–384

Khlebovich VV, Velikanov VP (1982) Osmotic and ionic regulation in green toad (*Bufo viridis* Laur.) from Sarikamish Lake. Sov J Ecol 12:228–231

Kikuchi M, Chiba A, Aoki K (1989) Entrainment of the circadian rhythms in Japanese common newt by melatonin injections. Zool Sci 6:1215

Kikuchi Y, Gomi T, Kimura A, Kishi K (1992) A distribution and localization of serotonin immunoreactive cells in the tracheal epithelium of *Hynobius tokyoensis*. Zool Sci 9:1290

Kim HH Chi YD, Moon YW (1984) The ultrastructure of the cutaneous pigment cells in *Rana nigromaculata coreana* Okada. Korean J Zool 27:231–240

Kim Y, Carpenter AM, Gregg KJ, Shahnaz Z, Carr JA (1995) Diurnal variation in α-melanocyte-stimulating hormone content of various brain regions and plasma of the Texas toad, *Bufo speciosus*. Gen Comp Endocrinol 98:50–56

King RB, Hauff S, Phillips JB (1994) Physiological color change in the green treefrog: responses to background brightness and temperature. Copeia 1994:422–432

Kirschner LB (1983) Sodium chloride absorption across the body surface: frog skins and other epithelia. Am J Physiol 244:R429–R443

Kirschner LB, Kerstetter T, Porter D, Alvarado RH (1971) Adaptation of larval *Ambystoma tigrinum* to concentrated environments. Am J Physiol 220:1814–1819

Klewen R (1985) Untersuchungen zur Ökologie und Populations-biologie des Feuersalamanders (*Salamandra salamandra terrestris* Lacepede 1788) an einer isolierten Population im Kreise Paderborn. Abh Westfal Mus Naturkd 47(1):1–51

Klewen R (1986) Population ecology of *Salamandra salamandra terrestris* in an isolated habitat. In: Roček Z (ed) Studies in herpetology Charles University, Prague, pp 395–398

Kloas W (1992) The role of atrial natriuretic peptide (ANP) in Amphibia. In: Proc Int Symp Amphibian Endocrinology, Tokyo, pp 11–12

Kobelt F, Linsenmair KE (1986) Adaptations of the reed frog *Hyperolius viridiflavus* (Amphibia, Anura, Hyperoliidae) to its arid environment. I. The skin of *Hyperolius viridiflavus nitidulus* in wet and dry season conditions. Oecologia 68:533–541

Kobelt F, Linsenmair KE (1992) Adaptations of the reed frog *Hyperlius viridiflavus*. (Amphibia: Anura: Hyperoliidae) to its arid environment. VI. The iridiophores in the skin as radiation reflectors. J Comp Physiol [B] 162:314–326

Kobelt F, Linsenmair KE (1995) Adaptations of the reed frog *Hyperlius viridiflavus* (Amphibia: Anura: Hyperoliidae) to its arid environment. VII. The heat budget of *Hyperolius viridiflavus nitidulus* and the evolution of an optimized body shape. J Comp Physiol [B] 165:110–124

Korf HW, Oksche A (1986) The pineal organ. In: Pang PKT, Schreibman MP (eds) Vertebrate endocrinology: fundamental and biomedical implications, vol 1. Morphological considerations. Academic Press, New York, pp 105–145

Krakauer T (1970) Tolerance limits of the toad, *Bufo marinus*, in south Florida. Comp Biochem Physiol 33:15–26

Krupa JJ (1994) Breeding biology of the great plains toad in Oklahoma. J Herpetol 28:217–224

Kühn ER, Darras VM, Gevaerts H (1985) Circadian and annual hormonal rhythms in amphibians. In: Follet BK, Chandola A (eds) The endocrine system and the environment. Springer, Berlin Heidelberg New York, pp 55–69

Kuzmin SL (1984) Age changes of feeding in *Hynobius keyserlingii* (Amphibia, Hynobiidae) (in Russian, English Summary). Zool Zh 63:1055–1060

Kuzmin SL (1991) The ecology and evolution of the amphibian cannibalism. J Bengal Nat Hist Soc (NS) 10:11–27

Kuzmin SL (1995) The problem of food competition in amphibians. Herpetol J 5:252–256

Kuzmin SL, Tarkhnishvili DN (1987) Age dynamics of feeding of sympatric Caucasian newts. Zool Zh 66:244–258 (in Russian, English Summary)

Lambert MRK (1984) Amphibians and reptiles. In: Cloudsley-Thompson JL (ed) Sahara desert. Pergamon Press, Oxford, pp 205–227

Lannoo MJ, Bachmann MD (1984) Aspects of cannibalistic morphs in a population of *Ambystoma t. tigrinum* larvae. Am Midl Nat 112:103–109

Lannoo MJ, Lowcock L, Bogart JP (1989) Sibling cannibalism in noncannibal morph *Ambystoma tigrinum* larvae and its correlation with high growth rates and early metamorphosis. Can J Zool 67:1911–1914

Larras-Regard E, Taurog A, Dorris M (1981) Plasma T_4 and T_3 levels in *Ambystoma tigrinum* at various stages of metamorphosis. Gen Comp Endocrinol 43:443–450

Larsen LO (1976) Physiology of molting. In: Lofts B (ed) Physiology of the Amphibia, vol III. Academic Press, New York, pp 53–99

Lascano EF, Segura ET (1971) Estudio morfologico de la piel del sapo adulto. Diferencias sexuales y variaciones anuales. Rev Soc Argent Biol 47:6–16

Lasiewski RC, Bartholomew GA (1969) Condensation as a mechanism for water gain in nocturnal desert poikilotherms. Copeia 1969:405–407

Laurent RF (1964) Adaptive modifications in frogs of an isolated highland in central Africa. Evolution 18:458–468

Lawler SP, Morin PJ (1993) Temporal overlap, competition, and priority effects in larval anurans. Ecology 74:174–182

Le Furgey A, Tisher CC (1981) Time course of vasopressin-induced formation of microvilli in granular cells of toad urinary bladder. J Membr Biol 61:13–19

Le Quang Trong NY (1967) Histogènes et histochimie des glandes cutanées de l'axolotl (*Ambystoma tigrinum* Green). Arch Zool Exp Gen 108:49–73

Lee AK (1967) Studies in Australian Amphibia. II. Taxonomy, ecology and evolution of the genus *Heleioporus* Gray (Anura: Leptodactylidae). Aust J Zool 15:367–439

Lee AK (1968) Water economy of the burrowing frog, *Heleioporus eyrei* (Gray). Copeia 1968:741–745

Lee AK, Mercer EH (1967) Cocoon surrounding desert-dwelling frogs. Science 157:87–88

Lee AR, Silove M, Katz U, Balinsky JB (1982) Urea cycle enzymes and glutamate dehydrogenase in *Xenopus laevis* and *Bufo viridis* adapted to high salinity. J Exp Zool 221:169–172

Leff LG, Bachmann MD (1986) Ontogenetic changes in predatory behavior of larval tiger salamanders (*Ambystoma tigrinum*). Can J Zool 64:1337–1344

Leips J, Travis J (1994) Metamorphic responses to changing food levels in two species of hylid frogs. Ecology 75:1345–1356

Levine SD, Franki N, Einhorn R, Hays RM (1976) Vasopressin-stimulated movement of drugs and uric acid across the toad urinary bladder. Kidney Int 9:30–35

Leviton AE, Anderson SC, Adler K, Minton SA (1992) Handbook to Middle East amphibians and reptiles. Society for the Study of Amphibians and Reptiles, Lawrence, Kansas

Lewinson D, Rosenberg M, Warburg MR (1982) Mitochondria-rich cells in salamander larva epidermis; ultrastructural descriptions and carbonic anhydrase activity. Biol Cell 46:75–84

Lewinson D, Rosenberg M, Warburg MR (1983) Developmental changes in the epidermal surface cells of salamander larva. Acta Zool (Stockh) 64:191–197

Lewinson D, Rosenberg M, Warburg MR (1984) "Chloride-cell"-like mitochondria-rich cells of salamander larva gill epithelium. Experientia 40:956–958

Lewinson D, Rosenberg M, Warburg MR (1987a) Ultrastructural and ultracytochemical studies of the gill epithelium in the larvae of *Salamandra salamandra* (Amphibia, Urodela). Zoomorphology 107:17–25

Lewinson D, Rosenberg M, Goldenberg S, Warburg MR (1987b) Carbonic anhydrase cytochemistry in mitochondria-rich cells of salamander larvae gill epithelium as related to age and H^+ and Na^+ concentrations. J Cell Physiol 130:125–132

Licht LE (1969) Observations on the courtship behavior of *Ambystoma gracile*. Herpetologica 25:49–52

Licht P, Feder MF, Bledsoe S (1975) Salinity tolerance and osmoregulation in the salamander *Batrachoseps*. J Comp Physiol 102:123–134

Liggins GW, Grigg GC (1985) Osmoregulation of the cane toad, *Bufo marinus*, in salt water. Comp Biochem Physiol [A] 82:613–619

Lillywhite HB, Licht P (1974) Movement of water over toad skin: functional role of epidermal sculpturing. Copeia 1974:165–171

Lillywhite HB, Licht P (1975) A comparative study of integumentary mucous secretions in amphibians. Comp Biochem Physiol [A] 51:937–941

Lillywhite HB, Licht P, Chelgren P (1973) The role of behavioral thermoregulation in the growth energetics of the toad, *Bufo boreas*. Ecology 54:375–383

Lindemann B, Voûte C (1976) Structure and function of the epidermis. In: Llinas R, Precht W (eds) Frog neurobiology. Springer, Berlin Heidelberg New York, pp 169–210

Lindgren E, Main AR (1961) Natural history notes from Jigalong. West Aust Nat 7:193–201

Lindquist SB, Bachmann MD (1980) Feeding behavior of the tiger salamander, *Ambystoma tigrinum*. Herpetologica 36:144–158

Lindquist SB, Bachmann MD (1982) The role of visual and olfactory cues in the prey catching behavior of the tiger salamander, *Ambystoma tigrinum*. Copeia 1982:81–90

Littlejohn MJ (1966) Amphibians of the Victorian mallee. Proc R Soc Victoria 79:597–604

Liu CC (1950) Amphibians of Western China. In: Schmidt PS, Ross LA (eds) Fieldiana: zoology memoirs, vol 2. Chicago Nat Hist Mus, Chicago, pp 1–396

Lodi G (1971) Histoenzymologic characterization of the flask cells in the skin of the crested newt under normal and experimental conditions. Atti Acc Sci (Torino) 105:561–570

Loeb MLG, Collins JP, Maret TJ (1994) The role of prey in controlling experession of a trophic polymorphism in *Ambystoma tigrinum nebulosum*. Funct Ecol 8:151–158

Lofts B (ed) (1976) Biology of the Amphibia, vols I–III. Academic Press, New York

Long DR (1987) Reproductive and lipid patterns of a semiarid adapted-anuran, *Bufo cognatus*. Tex J Sci 39:3–13

Loretz CA, Bern HA (1982) Prolactin and osmoregulation in vertebrates. Neuroendocrinology 35:292–304

Loveridge JP (1970) Observations on nitrogenous excretion and water relations of *Chiromantis xerampelina* (Amphibia, Anura). Arnoldia 5:1–6

Loveridge JP (1976) Strategies of water conservation in southern African frogs. Zool Afr 11:319–333

Loveridge JP, Craye G (1979) Cocoon fromation in two speices of Southern African frogs. S Afr J Sci 75:18–20

Loveridge JP, Withers PC (1981) Metabolism and water balance of active and cocooned African bullfrogs *Pyxicephalus adspersus*. Physiol Zool 54:203–214

Low BS (1976) The evolution of amphibian life histories in the desert. In: Goodall DW (ed) Evolution of desert biota. University of Texas Press, Austin, pp 149–195

Lucas EA, Reynolds WA (1967) Temperature selection by amphibian larvae. Physiol Zool 40:159–171

Luthardt G, Roth G (1979a) The influence of prey experience on movement pattern preference in *Salamandra salamandra* (L.). Z Tierpsychol 51:252–259

Luthardt G, Roth G (1979b) The relationship between stimulus orientation and stimulus movement pattern in the prey catching behavior of *Salamandra salamandra*. Copeia 1979:442–447

Luthardt G, Roth G (1983) The interaction of the visual and the olfactory systems in guiding prey catching behaviour in *Salamandra salamandra* (L.). Behaviour 83:69–79

Mack G, Hanke W (1977a) Studies on anuran osmomineral regulation. I. Comparison of the reaction to desiccation in different anuran species. Zool Jahrb Physiol 81:112–129

Mack G, Hanke W (1977b) Studies on anuran osmomineral regulation. II. Comparison of the effects of octopeptides in different anuran species. Zool Jahrb Physiol 81:177–190

MacMahon JA (1985) Deserts. The Audubon Society Nature Guides. Knopf Press, New York, 638 pp

Mahony MJ, Roberts JD (1986) Two new species of desert burrowing frogs of the genus *Neobatrachus* (Anura: Myobatrachidae) from Western Australia. Rec West Aust Mus 13:155–170

Main AR (1965) Frogs of southern Western Australia. Handbook no 8. Western Aust Nat Club, Perth, 73 pp

Main AR (1968) Ecology, systematics and evolution of Australian frogs. Adv Ecol Res 5:37–86

Main AR, Bentley PJ (1964) Water relations of Australian burrowing frogs and tree frogs. Ecology 45:379–382

Main AR, Littlejohn MJ, Lee AK (1959) Ecology of Australian frogs. Monogr Biol 8:396–411

Maina JN (1989) The morphology of the lung of the East African tree frog *Chirmantis petersi* with observations on the skin and the buccal cavity as secondary gas exchange organs. A TEM and SEM study. J Anat 165:29–43

Mainoya JR, Howell KM (1976) The occurrence of a mucopolysccharide layer in the integument of some Tanzanian anuran species. Univ Sci J (Dar Es Salaam Univ) 2:43–55

Maiorana VC (1976) Size and environmental predictability for salamanders. Evolution 30:599–613

Maiorana VC (1977) Observations of salamanders (Amphibia, Urodela, Plethodonthidae) dying in the field. J Herpetol 11:1–5

Maiorana VC (1978) Behavior of an unobservable species: diet selection by a salamander. Copeia 1978:664–672

Maloyi GMO (ed) (1979) Comparative physiology of osmoregulation in animals, vol 1. Academic Press, London

Malvin GM, Wood SC (1991) Behavioral thermoregulation of the toad, *Bufo marinus*: effects of air humidity. J Exp Zool 258:322–326

Malvin GM, Hood L, Sanchez M (1992) Regulation of blood flow through ventral pelvic skin by environmental water and NaCl in the toad *Bufo woodhousei*. Physiol Zool 65:540–553

Manteuffel G, Himstedt W (1986) Die Veränderung der Empfindlichkeit für Helligkeitskontraste während der Metamorphose von *Salamandra salamandra* (L.). Zool Jahrb Physiol 90:349–357

Manteuffel G, Wess O, Himstedt W (1977) Messungem am dioptrischen Apparat von Amphibienaugen und Berechnung der Sehschärfe in Wasser und Luft. Zool Jahrb Physiol 81:395–406

Marangio MS, Anderson JD (1977) Soil moisture preference and water relations of the marbled salamander, *Ambystoma opacum* (Amphibia, Urodela, Ambystomatidae). J Herpetol 11:169–176

Marshall E, Grigg GC (1980) Lack of metabolic acclimation to different thermal histories by tadpoles of *Limnodynastes peroni* (Anura: Leptodactylidae). Physiol Zool 53:1–7

Marshall EK, Smith HW (1930) The glomerular development of the vertebrate kidney in relation to habitat. Biol Bull 59:135–153

Martin AA (1967) Australian anuran life histories: some evolutionary and ecological aspects. In: Weatherly AH (ed) Australian inland waters and their fauna. Australian National University Press, Canberra, pp 175–191

Martin DL, Jaeger RG, Labat CP (1986) Territoriality in an *Ambystoma* salamander? Support for the null hypothesis. Copeia 1986:725–730

Martof BS (1962a) The behavior of Fowler's toad under various conditions of light and temperature. Physiol Zool 35:38–46

Martof BS (1962b) Some observations on the role of olfaction among salientian Amphibia. Physiol Zool 35:270–272

Mason JR, Stevens DA (1981) Discrimination and generalization among reagent grade odorants by tiger salamanders (*Ambystoma tigrinum*), Physiol Behav 26:647–653

Mason JR, Meredith M, Stevens DA (1981) Odorant discrimination by tiger salamanders after combined olfactory and vomeronasal nerve cuts. Physiol Behav 27:125–132

Mason JR, Stevens DA, Rabin MD (1980) Instrumentally conditioned avoidance by tiger salamanders (*Ambystoma tigrinum*) to reagent grade odorants. Chem Sense 5:99–105

Masoni A, Garcia-Romeu F (1979) Moulting in *Rana esculenta*: development of mitochondria-rich cells, morphological changes of the epithelium and sodium transport. Cell Thissue Res 197:23–38

Matsuda K, Toyoda F, Hayashi H, Kikuyama S (1994) Male-attractant in oviduct of the female newts, *Cynops pyrrhogaster*. Zool Sci 11 Suppl:16

Mayhew WW (1962) *Scaphiopus couchi* in California's Colorado desert. Herpetologica 18:153–161

Mayhew WW (1965) Adaptations of the amphibians, *Scaphiopus couchi*, to desert conditions. Am Midl Nat 74:95–109

Mayhew WW (1968) Biology of desert amphibians and reptiles. In: Brown GW (ed) Desert biology, vol 1. Academic Press, New York, pp 195–256

McClanahan L (1964) Osmotic tolerance of the muscles of two desert-inhabiting toads, *Bufo cognatus* and *Scaphiopus couchi*. Comp Biochem Physiol 12:501–508

McClanahan LL (1967) Adaptations of the spadefoot toad, *Scaphiopus couchi*, to desert environments. Comp Biochem Physiol 20:73–99

McClanahan LL (1972) Changes in body fluids of burrowed spadefoot toads as a function of soil water potential. Copeia 1972:209–216

McClanahan LL (1975) Nitrogen excretion in arid-adapted amphibians. In: Hadley NF (ed) Environmental physiology of desert organisms. Dowden Hutchinson and Ross, Stroudsbery, pp 106–116

McClanahan LL, Baldwin R (1969) Rate of water uptake through the integument of the desert toad, *Bufo punctatus*. Comp Biochem Physiol 28:381–389

McClanahan LL, Shoemaker VH (1987) Behavior and thermal relations of the arboreal frog *Phyllomedusa sauvagei*. Nat Geogr Res 3:11–21

McClanahan LL, Shoemaker VH, Ruibal R (1976) Structure and function of the cocoon of a ceratophrid frog. Copeia 1976:179–185

McClanahan LL, Stinner JN, Shoemaker VH (1978) Skin lipids, water loss, and energy metabolism in a South American tree frog (*Phyllomedusa sauvagei*). Physiol Zool 51:179–187

McClanahan LL, Ruibal R, Shoemaker VH (1983) Rate of cocoon formation and its physiological correlates in a ceratophryd frog. Physiol Zool 56:430–435

McClanahan LL, Ruibal R, Shoemaker VH (1994) Frogs and toads in deserts. Sci Am 271:64–70

McDiarmid RW, Foster MS (1987) Cocoon formation in another hylid frog, *Smilisca baudini*. J Herpetol 21:352–355

McGregor JH (1989) Olfaction as an orientation mechanism in migrating *Ambrstoma mayculatum*. Copeia 1989:779–781

Meek R (1983) Body temperature of two species of desert amphibians, *Rana perezi* and *Bufo mauritanicus*. Br J Herpetol 6:284–286

Meissner K (1970) Zur atypischen Struktur, Dynamic und Aktualgenese des grabspezifischen Appetenzverhalten der Knoblauchkröte (*Pelobates f. fuscus* Laur.; Pelobatidae, Anura). Biol Zentralbl 89:409–434

Mellado J, Dakki M (1988) Inventaire commente des amphibiens et reptiles. Bull Inst Sci Rabat 12:171–181

Mellado J, Mateo JA (1992) New records of Moroccan herpetofauna. Herpetol J 2:58–61

Mellado J, Allabou A, Alaoui B (1988) L'herpétofaune du project de Parc National du Massa (Agadir, Maroc): un aperçu écologique et ses implications dans le développement du plan d'aménagement. Acta Oecol 9:55–74

Mia AJ, Oakford LX, Torres L, Herman C, Yorio T (1987) Morphometric analysis of epithelial cells of frog urinary bladder. I. Effect of antidiuretic hormone, calcium ionophore (A23187) and PGE$_2$. Tissue Cell 19:437–450

Michael MI, Yacob AY (1974) The development, growth and degeneration of the pronephric system in anuran amphibians of Iraq. J Zool Lond 174:407–417

Michel G, Ouedraogo Y, Chauvet J, Katz U, Acher R (1993) Differential processing of provasotocin: relative increase of Hydrin 2 (Vasotocinyl-Gly) in amphibians able to adapt to an arid environment. Neuropeptides 25:139–143

Middler SA, Kleeman CR, Edwards E (1968) The role of the urinary bladder in salt and water metabolism of the toad, *Bufo marinus*. Comp Biochem Physiol 26:57–68

Middler SA, Kleeman CR, Edwards E, Brody D (1969) Effect of adenohypophysectomy on salt and water metabolism of the toad *Bufo marinus* with studies on hormonal replacement. Gen Comp Endocrinol 12:290–304

Miller MR, Robbins ME (1954) The reproductive cycle in *Taricha torosa* (*Triturus torosus*). J Exp Zool 215:415–440

Miller MA, Stebbins R (1964) The life of desert animals in Joshua Tree National Monument. University of California Press, Berkeley, 452 pp

Moore JA (1964) Physiology of Amphibia. Academic Press, New York

Moore RG, Moore BA (1980) Observations on the body temperature and activity in the red spotted toad, *Bufo punctatus*. Copeia 1980:362–363

Morafka DJ (1977) A biogeographical analysis of the Chihuahuan desert through its herpetofauna. Junk Press, The Hague, 313 pp

Morin PJ, Lawler SP, Johnson EA (1988) Competition between aquatic insects and vertebrates: interaction strength and higher order interactions. Ecology 69:1401–1409

Moriya T (1982) Prolactin induces increase in the specific gravity of salamander, *Hynobius retardatus*, that raises adaptability to water. J Exp Zool 223:83–88

Moriya T, Dent JN (1986) Hormonal interaction in the mechanism of migratory movement in the newt, *Notophthalmus viridescens*. Zool Sci 3:669–676

Mullen TL, Alvarado RH (1976) Osmotic and ionic regulation in amphibians. Physiol Zool 49:11–23

Mullens DP, Hutchison VH (1992) Diel, seasonal, postprandial and food-deprived thermoregulatory behaviour in tropical toads (*Bufo marinus*). J Therm Biol 17:63–67

Muller J, Kachadorian WA, DiScala VA (1980) Evidence that ADH-stimulated intramembrane particle aggregates are transferred from cytoplasmic to luminal membranes in toad bladder epithelial cells. J Cell Biol 85:83–95

Mushinsky HR, Brodie ED (1975) Selection of substrate pH by salamanders. Am Midl Nat 93:440–443

Nagel W, Katz U (1991) The effect of aldosterone on sodium transport and membrane conductances in toad skin (*Bufo viridis*). Pflügers Arch 418:319–324

Nakashima H, Kamishima Y (1990) Regulation of water permeability of the skin of the treefrog, *Hyla arborea japonica*. Zool Sci 7:371–376

Netchitailo P, Feuilloley M, Pelletier G, Cantin M, De Lean A, Leboulenger F, Vaudry H (1986) Localization and characterization of atrial natriuretic factor (ANF) -like peptide in the frog atrium. Peptides 7:573–579

Netchitatilo P, Leboulenger F, Cantin M, Gutkowska J, Vaudry H (1987) Atrial natriuretic factor-like immunoreactivity in the central nervous system of the frog. Neuroscience 22:341–360

Netchitailo P, De Lean A, Ong H, Cantin M, Gutkowska J, Leboulenger F, Vaudry H (1988) Localization and identification of immunoreactive atrial natriuretic factor (ANF) in the frog ventricle. Peptides 9:1–6

Newman RA (1987) Effects of density and predation on *Scaphiopus couchi* tadpoles in desert ponds. Oecologia 71:301–307

Newman RA (1988a) Genetic variation for larval anuran (*Scaphiopus couchii*) development time in an uncertain environment. Evolution 42:763–773

Newman RA (1988b) Adaptive plasticity in development of *Scaphiopus couchii* tadpoles in desert ponds. Evolution 42:774–783

Newman RA (1989) Developmental plasticity of *Scaphiopus couchi* tadpoles in an unpredictable environment. Ecology 70:1775–1787

Newman RA (1994) Effects of changing density and food level on metamorphosis of a desert amphibian, *Scaphiopus couchii*. Ecology 75:1085–1096

Newman RA, Dunham AE (1994) Size at metamorphosis and water loss in a desert anuran (*Scaphiopus couchii*). Copeia 1994:372–381

Nicholas JS (1922) The reactions of *Amblystoma tigrinum* to olfactory stimuli. Anat Rec 20:257–281

Nielsen HI (1978a) Ultrastructural changes in the dermal chromatophore unit of *Hyla arborea* during color change. Cell Tissue Res 194:405–418

Nielsen HI (1978b) The effect of stress and adrenaline on the color of *Hyla cinerea* and *Hyla arborea*. Gen Comp Endocrinol 36:543–552

Nielsen HI, Bereiter-Hahn J (1982) Hormone induced chromatophore changes in the European tree frog, *Hyla arborea*, in vitro. J Zool (Lond) 198:363–381

Nishimura H (1987) Role of the renin-angiotensin system in osmoregulation. In: Pang PKT, Schreibman MP (eds) Fundamental and biomedical implications, vol 2. Academic Press, New York, pp 157–187

Noble K (1931) The Biology of the Amphibia. Dover, Mineola, 577 pp

Nolly H, Fasciolo JC (1971a) The renin-angiotensin system in *Bufo arenarum* and *Bufo paracnemis*. Comp Biochem Physiol [A] 39:823–831

Nolly H, Fasciolo JC (1971b) Renin-angiotensin system and sodium homeostasis in *Bufo arenarum*. Comp Biochem Physiol [A] 39:833–841

Norris DO, Austin HB, Hijazi AS (1989) Induction of cloacal and dermal glands of tiger salamander larvae, (*Ambystoma tigrinum*): effects of testosterone and prolactin. Gen Comp Endocrinol 73:194–204

Norris DO, Jones RE, Criley BB (1973) Pituitary prolactin levels in larval, neotenic and metamorphosed salamanders (*Ambystoma tigrinum*). Gen Comp Endocrinol 20:437–442

Nouwen EJ, Kühn ER (1983) Radioimmunoassay of arginine vasotocin and mesotocin in serum of the frog *Rana ridibunda*. Gen Comp Endocrinol 50:242–251

Nouwen EJ, Kühn ER (1985) Volumetric control of arginine vasotocin and mesotocin release in the frog (*Rana ridibunda*). J Endocrinol 105:371–377

Nussbaum M (1886) Zur Kenntnis der Nierenorgane. Arch Mikrosk Anat 27:442–480

Nyman S, Wilkinson RF, Hutcherson JE (1993) Cannibalism and size relations in a cohort of larval ringed salamanders (*Ambystoma annulatum*). J Herpetol 27:78–84

Odendaal FJ, Bull CM, Nias RC (1982) Habitat selection in tadpoles of *Ranidella signifera* and *R. riparia* (Anura: Leptodacylidae). Oecologia 52:411–414

Oguchi A, Mita M, Ohkawa M, Kawamura K, Kikuyama S (1994) Analysis of lung sufactant in the metamorphosing bullfrog (*Rana catesbeiana*). J Exp Zool 269:515–521

O'Hara RK, Blaustein AR (1982) Kin preference behavior in *Bufo boreas* tadpoles. Behav Ecol Sociobiol 11:43–49

Ohkawa M, Kawamura K, Kikuyama S (1989) Lung maturation in bullfrog tadpoles during metamorphosis. Zool Sci 6:1195

Oldham RS (1966) Spring movements in the American toad. *Bufo americanus*. Can J Zool 44:63–100

Olivereau M, Olivereau JM, Kikuyama S, Yamamoto K (1990) Hypothalamo-hypophysial axis osmoregulation. Fortschr Zool 38:371–383

Orgeig S, Daniels CB, Smits AW (1994) The composition and function of the pulmonary surfactant system during metamorphosis in the tiger salamander *Ambystoma tigrinum*. J Comp Physiol [B] 164:337–342

Orlova VF, Uteshev VK (1986) The tetraploid toad of the *Bufo viridis* group from Dzungurian Gobi, Mongolia. In: Rocek Z (ed) Studies in herpetology. Charles University, Prague, pp 143–146

Overton E (1904) Neununddreissig Thesem über die Wasserökonomie der Amphibien und die osmotischen Eigenschaften der Amphibienhaut. Verh Phys Med Ges Würzb 36:277–295

Packard GC (1976) Devonian amphibians: did they excrete carbon dioxide via skin, gills, or lungs? Evolution 30:270–280

Packer WC (1963) Dehydration, hydration, and burrowing berhavior in *Heleioporus eyrei* (Gray) (Leptodactylidae). Ecology 44:643–651

Packer WC (1966) Embryonic and larval development of *Heleioporus eyrei* (Amphibia: Leptodactylidae). Copeia 1966:92–97

Pandian TJ, Marian MP (1985) Time and energy costs of metamorphosis in the Indian bullfrog *Rana tigrina*. Copeia 1985:653–662

Pang PKT (1977) Osmoregulatory functions of neurohypophysial hormones in fishes and amphibians. Am Zool 17:739–749

Pang PKT, Sawyer WH (1978) Renal and vascular responses of the bullfrog (*Rana catesbeiana*) to mesotocin. Am J Physiol 235:F151–F155

Pang PKT, Schreibman MP (eds) (1987) Vertebrate endocrinology: fundamentals and biomedical implications, vols 1, 2. Academic Press, New York

Pang PKT, Furspan PB, Sawyer WH (1983) Evolution of neurohypophyseal hormone actions in vertebrates. Am Zool 23:655–662

Parakkal PF, Matoltsy AG (1964) A study of fine structure of the epidermis of *Rana pipiens*. J Cell Biol 20:85–94

Parker HW (1940) The Australasian frogs of the family Leptodactylidae. Novit Zool 42(1):1–216

Pashkova IM (1985) Seasonal changes in the heat resistance of *Bufo viridis* and its muscles and of contractile muscle models. J Therm Biol 10:105–108

Passmore NI (1972) Integrading between members of the "regularis group" of toads in South Africa. J Zool (Lond) 167:143–151

Pattle RE (1976) The lung surfactant in the evolutionary tree. In: Hughes GM (ed) Respiration of amphibious vertebrates. Academic Press, New York, pp 233–255

Pattle RE, Schock C, Creasey JM, Hughes GM (1977) Surpellic films, lung surfactant, and their cellular origin in newt, caecilian, and frog. J Zool (Lond) 182:125–136

Paulson BK, Hutchison VH (1987) Blood changes in *Bufo cognatus* following acute heat stress. Comp Biochem Physiol [A] 87:461–466

Pearl M, Taylor A (1985) Role of cytoskeleton in the control of transcellular water flow by vasopressin in amphibian urinary bladder. Biol Cell 55:163–172

Pearman PB (1991) Effects of habitat size on tadpole populations. Ecology 74:1982–1991

Pearman PB (1995) Effects of pond size and consequent predator density on two species of tadpoles. Oecologia 102:1–8

Pearson PG (1955) Population ecology of the spadefoot toad, *Scaphiopus h. holbrooki* (Harlan). Ecol Monogr 25:233–267

Petranka JW (1982) Geographic variation in the mode of reproduction and larval characteristics of the small-mouthed salamander (*Ambystoma texanum*) in the east-central United States. Herpetologica 38:475–485

Petranka JW (1984a) Breeding migrations, breeding season, clutch size, and oviposition of stream-breeding *Ambystoma texanum*. J Herpetol 18:106–112

Petranka JW (1984b) Incubation, larval growth, and embryonic and larval survivorship of small-mouth salamanders (*Ambystoma texanum*) in streams. Copeia 1984:862–868

Petranka JW (1984c) Sources of interpopulational variation in growth responses of larval salamanders. Ecology 65:1857–1865

Petranka JW (1989) Density-dependent growth and survival of larval *Ambystoma*: evidence from whole-pond manipulation. Ecology 70:1752–1767

Petranka JW, Petranka JG (1980) Selected aspects of the larval ecology of the marbled salamander *Ambystoma opacum* in the southern portion of its range. Am Midl Nat 104:352–363

Petranka JW, Sih A (1986) Environmental instability, competition, and density-dependent growth and survivorship of a stream-dwelling salamander. Ecology 67:729–736

Petriella S, Reboreda JC, Otero M, Segura ET (1989) Antidiuretic responses to osmotic cutaneous stimulation in the toad, *Bufo arenarum*. J Comp Physiol [B] 159:91–95

Pfennig D (1990) The adaptive significance of an environmentally-cued developmental switch in an anuran tadpole. Oecologia 85:101–107

Pfennig DW (1992a) Polyphenism in spadefoot toad tadpoles as a locally adjusted evolutionary stable strategy. Evolution 46:1408–1420

Pfennig DW (1992b) Proximate and functional causes of polyphenism in an anuran tadpole. Funct Ecol 6:167–174

Pfennig DW, Collins JP (1993) Kinship affects morphogenesis in cannibalistic salamanders. Nature 362:836–838

Pfennig DW, Marby A, Orange D (1991) Environmental causes of correlations between age and size at metamorphosis in *Scaphiopus multiplicatus*. Ecology 72:2240–2248

Pfennig DW, Reeve HK, Sherman PW (1993) Kin recognition and cannibalism in spadefoot toad tadpoles. Anim Behav 46:87–94

Pfitzner W (1879) Die Leydig'schen Schleimzellen in der Epidermis der Larven von *Salamandra maculosa*. PhD dissertation, University of Kiel, pp 3–21

Pfitzner W (1880) Die Epidermis der Amphibien. Morphol Jahrb 6:469–526

Picheral B (1970) Les tissus élaborateurs d'hormones stéroides chez les amphibiens urodèles. Z Zellforsch 107:68–86

Pierce BA, Mitton JB, Rose FL (1981) Allozyme variation among large, small and cannibal morphs of the tiger salamander inhabiting the Llano Estacado of west Texas. Copeia 1981:590–595

Pierce BA, Mitton JB, Jacobson L, Rose FL (1983) Head shape and size in cannibal and noncannibal larvae of the tiger salamander from west Texas. Copeia 1983:1006–1012

Pilkington JB, Simkiss K (1966) The mobilization of the calcium carbonate deposits in the the endolymphatic sacs of metamorphosing frogs. J Exp Biol 45:329–341

Plasa L (1979) Heimfindeverhalten bei *Salamandra salamandra* (L.). Z Tierpsychol 51:113–125

Platt JE, Christopher MA (1977) Effects of prolactin on the water and sodium content of larval tissues from neotenic and metamorphosing *Ambystoma tigrinum*. Gen Comp Endocrinol 31:243–248

Platt JE, Brown GB, Erwin SA, McKinley KT (1986) Regression and acid phosphatase activity in metamorphosing *Ambystoma tigrinum*. Gen Comp Endocrinol 61:376–382

Pons G, Guardabassi A, Pattono P (1982) The kindney of *Hyla arborea* (L.) (Amphibia Hylidae) in autumn, winter and spring: histological and ultrastructural observations. Monit Zool Ital (NS) 16:261–281

Porter GA (1971) The action of aldosterone on transepithelial sodium transport in isolated ventral toad skin. Gen Comp Endocrinol 16:443–451

Pough FH, Wilson RE (1970) Natural daily temperature stress, dehydration and acclimation in juvenile *Ambystoma maculatum* (Shaw) (Amphibia: Caudata). Physiol Zool 43:194–205

Pough FH, Stewart MM, Thomas RG (1977) Physiological basis of habitat partitioning in a Jamaican *Eleutherodactylus*. Oecologia 27:285–293

Pough FH, Taigen TL, Stewart MM, Brussard PF (1983) Behavioral modification of evaporative water loss by Puerto Rican frog. Ecology 64:244–252

Poynton JC (1964) The Amphibia of southern Africa: a faunal study. Ann Natal Mus 17:1–334

Poynton JC, Pritchard S (1976) Notes on the biology of *Breviceps* (Anura: Microhylidae). Zool Afr 11:313–318

Preest MR (1993) Mechanisms of growth rate reduction in acid-exposed larval salamanders, *Ambystoma maculatum*. Physiol Zool 66:686–707

Preest MR, Pough FH (1989) Interaction of temperature and hydration on locomotion of toads. Funct Ecol 3:693–699

Preest MR, Brust DG, Wygoda ML (1992) Cutaneous water loss and the effects of temperature and hydration state on aerobic metabolism of canyon treefrogs, *Hyla arenicolor*. Herpetologica 48:210–219

Propper CR, Sasongko BWP, Hillyard SD, Proefrock K (1992) Immunohistochemical staining in the brain and ovary of desert toads using an antibody against the peptide angiotensin II. Proc Int Symp Amphibian Endocrinology, Tokyo, 15 pp

Pruett SJ, Hoyt DF, Stiffler DF (1991) The allometry of osmotic and ionic regulation in Amphibia with emphasis on intraspecific scaling in larval *Ambystoma tigrinum*. Physiol Zool 64:1173–1199

Przyrembel C, Keller B, Neumeyer C (1995) Trichromatic color vision in the salamander (*Salamandra salamandra*). J Comp Physiol [A] 176:575–586

Putnam RW, Bennett AF (1981) Thermal dependence of behavioural performance of anuran amphibians. Anim Behav 29:502–509

Putnam RW, Hillman SS (1977) Activity responses of anurans to dehydration. Copeia 1977:746–749

Rabb GB (1973) Evolutionary aspects of the reproductive behavior of frogs. In: Vial JL (ed) Evolutionary biology of the anurans. University of Missouri Press, Columbia, pp 213–227

Rapoport J, Kachadorian WA, Muller J, Franki N, Hays RM (1981) Stabilization of vasopressin-induced membrane events by bifunctional imidoester. J Cell Biol 89:261–266

Ray J (1970) Instrumental avoidance learning by the tiger salamander *Ambystoma tigrinum*. Anim Behav 18:73–77

Raymond LR, Hardy LM (1990) Demography of a population of *Ambystoma talpoideum* (Caudata: Ambystomatidae) in Northwestern Louisiana. Herpetologica 46:371–382

Reed CSA, Marx H (1959) A herpetological collection from northeastern Iraq. Trans Kansas Acad Sci 62:91–122

Reichling H (1958) Transpiration und Vorzugstemperatur mitteleuropäischer Reptilien und Amphibien. Zool Jahrb (Physiol) 67:1–64

Reno HW, Gehlbach FR, Turner RA (1972) Skin and aestivational cocoon of the aquatic amphibian, *Siren intermedia* Le Conte. Copeia 1972:625–631

Richards CA (1982) The alteration of chromatophore expression by sex hormones in the Kenyan reed frog, *Hyperolius viridiflavus*. Gen Comp Endocrinol 46:59–67

Richmond ND (1947) Life history of *Scaphiopus holbrooki holbrooki* (Harlan). I. Larval development and behavior. Ecology 28:53–67

Rick R, Dorge A, Katz U, Bauer R, Thurau K (1980) The osmotic behaviour of toad skin epithelium (*Bufo viridis*). Pfluegers Arch 385:1–10

Rivero-Blanco C, Dixon JR (1979) Origin and distribution of the herpetofauna of the dry lowland regions of northern South America. Monogr Mus Nat Hist Kansas 7:281–295

Roberts JD (1981) Terrestrial breeding in the Australian leptodactylid frog *Myobatrachus gouldii* (Gray). Aust Wildl Res 8:451–462

Roberts JD (1984) Terrestrial egg deposition and direct development in *Arenophryne rotunda* Tyler, a myobatrachid frog from coastal sand dunes at Shark Bay, W.A. Aust Wildl Res 11:191–200

Roberts JD (1985) Population density estimates for *Arenophryne rotunda*: is the round frog rare? In: Grigg G, Shine R, Ehmann H (eds) Biology of Australasian frogs and reptiles. Royal Zoological Society, NSW, Sydney, pp 463–467

Roberts JD (1989) The biology of *Arenophryne rotunda* (Anura: Myobatrachidae): a burrowing frog from Shark Bay, Western Australia. Research in Shark Bay, pp 287–297

Roberts JD, Mahony M, Kendrick P, Majors CM (1991) A new species of burrowing frog, *Neobatrachus* (Anura: Myobatrachidae), from the eastern wheatbelt of Western Australia. Rec West Aust Mus 15:23–32

Robertson DR (1971a) Cytological and physiological activity of the ultimobranchial glands in the premetamorphic anuran *Rana catesbeiana*. Gen Comp Endocrinol 16:329–341

Robertson DR (1971b) Endocrinology of amphibian ultimobranchial glands. J Exp Zool 178:101–114

Robertson DR (1971c) Calcitonin in amphibians and the relationship of the paravertebral lime sacs with carbonic anhydrase. In: Calcium, Parathyroid hormone and the calcitonins. Proc 4th Parathyroid Conf, Chapel Hill, Noth Carolina, Exerpta Medica, Ansterdam, pp 21–28

Robertson DR (1987) The ultimobranchial body. In: Pang PKT, Schreibman MP (eds) Vertebrate endocrinology: fundamentals and biomedical implications, vol 1. Academic Press, New York, pp 235–259

Robertson JD (1957) The habitat of the early vertebrates. Biol Rev 32:156–187

Rohrbach JW, Stiffler DF (1987) Blood-gas, acid-base, and electrolyte responses to exercise in larval *Ambystoma tigrinum*. J Exp Zool 244:39–47

Romspert AP, McClanahan LL (1981) Osmoregulation of the terrestrial salamander, *Ambystoma tigrinum*, in hypersaline media. Copeia 1981:400–405

Rose FL, Armentrout D (1974) Population estimates of *Ambystoma tigrinum* inhabiting two playa lakes. J Anim Ecol 43:671–679

Rose FL, Armentrout D (1976) Adaptive strategies of *Ambystoma tigrinum* Green inhabiting the Llano Estacado of West Texas. J Anim Ecol 45:713–729

Rose SM (1960) A feedback mechanism of growth control in tadpoles. Ecology 41:188–196

Rose W (1962) South African reptiles and amphibians. Maskew Miller, Cape Town, 494 pp

Rosenberg M, Warburg MR (1991) The evidence for Na^+, K^+-ATPase activity in the epidermis of *Pelobates syriacus* tadpoles and toads. Biol Cell 71:281–287

Rosenberg M, Warburg MR (1992) Ultrastructure and histochemistry of ventral epidermis in *Pelobates syriacus* (Anura; Pelobatidae) tadpoles. Biol Struct Morphogen 4:164–170

Rosenberg M, Warburg MR (1993) The ventral epidermis of *Pelobates syriacus* (Anura; Pelobatidae). Isr J Zool 29:235–243

Rosenberg M, Lewinson D, Warburg MR (1982) Ultrastructural studies of the epidermal Leydig cell in larvae of *Salamandra salamandra* (Caudata, Salamandrida). J Morphol 174:275–281

Roth G, Luthardt G (1980) The role of early sensory experience in the prey catching responses of *Salamandra salamandra* to stationary prey. Z Tierpsychol 52:141–148

Roth JJ (1973) Vascular supply to the ventral pelvic region of anurans as related to water balance. J Morphol 140:443–460

Rouillé Y, Michel G, Chauvet MT, Chauvet J, Acher R (1989) Hydrins, hydroosmotic neurohypophysial peptides: osmoregulatory adaptation in amphibians through vasotocin precursor processing. Proc Natl Acad Sci USA 86:5272–5275

Rovedatti MG, Castane PM, Salibian A, Espina S (1988) Studies on the urinary nitrogen waste products in South American anurans from different habitats. Comp Biochem Physiol [A] 90:249–252

Ruibal R (1962) The adaptive value of bladder water in the toad, *Bufo cognatus*. Physiol Zool 35:218–223

Ruibal R, Hillman S (1981) Cocoon structure and function in the burrowing hylid frog, *Pternohyla fodiens*. J Herpetol 15:403–408

Ruibal R, Tevis L, Roig V (1969) The terrestrial ecology of the spadefoot toad *Scaphiopus hammondi*. Copeia 1969:571–584

Saidapur SK (1982) Structure and function of postovulatory follicles (corpora lutea) in the ovaries of nonmammalian vertebrates. Int Rev Cytol 75:243–285

Salibian A (1977) Transporte de cloro y de sodio a traves de la piel in situ de anfibios sudamericanos. Bol Mus Nac Hist Nat Chile 35:121–163

Salibian A, Fichera LE (1984) Excretion nitrogenada urinaria de *Bufo arenarum*. Bol Fisiol Anim Univ S Paulo 8:109–118

Salibian A, Pezzani-Hernandez S, Garcia Romeu F (1968) In vivo ionic exchange through the skin of the South American frog, *Leptodactylus ocellatus*. Comp Biochem Physiol 25:311–317

Salibian A, Preller A, Robres L (1971) In vivo ionic uptake through the skin of the South American toad *Bufo arunco*. Rev Can Biol 30:115–124

Salthe SN (1969) Reproductive modes and the number and sizes of ova in the urodeles. Am Midl Nat 82:467–490

Salthe SN, Duellman WE (1973) Quantitative constraints associated with reproductive mode in anurans. In: Vial JL (ed) Evolutionary biology of the anurans. University of Missouri Press, Columbia, pp 229–249

Salthe SN, Mecham JS (1974) Reproductive and courtship patterns. In: Lofts B (ed) Biology of the Amphibia, vol II. Academic Press, New York, pp 309–521

Samollow PB (1980) Selective mortality and reproduction in a natural population of *Bufo boreas*. Evolution 34:18–39

Sampson HW, Cannon MS, Davis RW (1987) The calcified amorphous layer of the skin of *Bufo marinus* (Amphibia: Anura). J Zool (Lond) 213:63–69

Sawyer WH (1956) The hormonal control of water and salt-electrolyte metabolism with special reference to the Amphibia. Mem Soc Endocrinol 5:44–59

Sawyer WH (1957) Increased renal reabsorption of osmotically free water by the toad (*Bufo marinus*) in response to neurohypophysial hormones. Am J Physiol 189:546–568

Sawyer WH (1972) Lungfishes and amphibians: endocrine adaptation and the transition from aquatic to terrestrial life. Fed Proc 31:1609–1614

Sawyer WH, Pang PKT (1975) Endocrine adaptation to osmotic requirements of the environment: endocrine factors in osmoregulation by lungfishes and amphibians. Gen Comp Endocrinol 25:224–229

Sawyer WH, Pang PKT (1987) Endocrine osmoregulation and vertebrate evolution. In: Pang PKT, Schreibman MP (eds) Vertebrate endocrinology: fundamentals and biomedical implications, vol 2. Regulation of water and electrolytes. Academic Press, New York, pp 293–305

Sawyer WH, Sawyer MK (1952) Adaptive responses to neurohypophyseal fractions in vertebrates. Physiol Zool 25:84–98

Sawyer WH, Schisgall RM (1956) Increased permeability of the frog bladder to water in response to dehydration and neurohypophysial extracts. Am J Physiol 187:312–314

Sawyer WH, Pang PKT, Galli-Gallardo SM (1978) Environment and osmoregulation among lungfishes and amphibians. In: Assenmacher I, Farner DS (eds) Environmental endocrinology. Springer, Berlin Heidelberg New York, pp 210–215

Scheer BT, Mumbach MW, Thompson AR (1974) Salt balance and osmoregulation in salientian amphibians. In: Florkin M, Scheer BT (eds) Chemical zoology, vol IX. Amphibia and Reptilia. Academic Press, New York, pp 51–65

Scheuermann DW, Adriaensen D, Timmermans JP, de Groodt-Lasseel MHA (1989) Neuroepithelial endocrine cells in the lung of *Ambystoma mexicanum*. Anat Rec 225:139–149

Schindelmeiser J, Schindelmeiser I, Greven H (1983) Hepatic arginase activity in intra- and extrauterine larvae of the ovoviviparous slamander, *Salamandra salamandra* (L.) (Amphibia, Urodela). Comp Biochem Physiol [B] 75:471–473

Schindelmeiser J, Greven H (1981) Nitrogen excretion in intra- and extrauterine larvae of the ovoviviparous salamander, *Salamandra salamandra* (L.) (Amphibia, Urodela). Comp Biochem Physiol 70 A:568–565

Schiotz A (1967) The tree frogs (Rhacophoridae) of West Africa. Spolia Zool Mus Haun 25:1–346

Schiotz A (1976) Zoogeographical patterns in the distribution of East Aftican treefrogs (Anura: Ranidae). Zool Afr 11:335–338

Schmajuk NA, Segura ET (1982) Behavioral regulation of water balance in the toad *Bufo arenarum*. Herpetologica 38:296–301

Schmid WD (1965a) High temperature tolerances of *Bufo hemiophrys* and *Bufo cognatus*. Ecology 46:559–560

Schmid WD (1965b) Some aspects of the water economies of nine species of amphibians. Ecology 46:261–269

Schmid WD (1968) Natural variations in nitrogen excretion of amphibians from different habitabs. Ecology 49:180–185

Schmid WD (1969) Physiological specialization of amphibians to habitats of varying aridity. In: Hoff CC, Riedsel ML (eds) Physiological systems in semiarid environments. University of New Mexico Press, Albuquerque, pp 135–142

Schmid WD, Barden RE (1965) Water permeability and lipid content of amphibian skin. Comp Biochem Physiol 15:423–427

Schmid WD, Underhill JC (1946) Sodium transport by skin of amphibian species from different habitats. Ecology 45:864–865

Schmuck R, Linsenmair KE (1988) Adaptations of the reed frog *Hyperolius viridiflavus* (Amphibia, Anura, Hyperoliidae) to its arid environment. III. Aspects of nitrogen metabolism and osmoregulation in the reed frog, *Hyperolius viridiflavus taeniatus*, with special reference to the role of iridiophores. Oecologia 75:354–361

Schmuck R, Kobelt F, Linsenmair KE (1988) Adaptations of the reed frog *Hyperolius viridiflavus* (Amphibia, Anura, Hyperoliidae) to its arid environment. V. Iridiophores and nitrogen metabolism. J Comp Physiol [B] 158:537–546

Schmuck R, Geise W, Linsenmair KE (1994) Life cycle strategies and physiological adjustments of reedfrog tadpoles (Amphibia, Anura, Hyperoliidae) in relation to environmental conditions. Copeia 1994:996–1007

Schneider CW (1968) Avoidance learning and the response tendencies of the larval salamander *Ambystoma punctatum* to photic stimulation. Anim Behav 16:492–495

Schneider H (1971) Die Steuerung des täglichen Rufbeginns beim Laubfrosch, *Hyla arborea arborea* (L.). Oecologia 8:310–320

Schneider H, Nevo E (1972) Bio-acoustic study of the yellow-lemon treefrog, *Hyla arborea savignyi* Audouin. Zool Jahrb Physiol 76:497–506

Schoffeniels E, Tercafs RR (1965 1966) L'osmorégulation chez les batraciens. Ann Soc R Zool Belg 96:23–39

Schultheiss H (1973) Der Einfluß von ACTH und Corticosteroiden auf den Stickstoff-Stoffwechsel während der Metamorphose des mexikanischen Axolotl (*Ambystoma mexicanum* Cope). Zool Jahrb Physiol 77:199–227

Schultheiss H (1977) The hormonal regulation of urea excretion in the Mexican axolotl (*Ambystoma mexicanum* Cope). Gen Comp Endocrinol 31:45–52

Schultheiss H, Hanke W (1978) Urea excretion rates and waste nitrogen concentrations during metamorphosis of *Xenopus laevis* Daudin. Comp Biochem Physiol [A] 61:567–570

Schwalbe G (1896) Zur Biologie und Entwicklungsgeschichte von *Salamandra atra* und *maculosa*. Z Biol (NF) 16:340–396

Schwalm PA, McNulty JA (1980) The morphology of dermal chromatophores in the infrared-reflecting glass-frog *Centrolenella fleischmanni*. J Morphol 163:37–44

Schwalm PA, Starrett PH, McDiarmid RW (1977) Infrared reflectance in leaf-sitting neotropical frogs. Science 196:1225–1227

Scott DE (1993) Timing of reproduction of paedomorphic and metamorphic *Ambystoma talpoideum*. Am Midl Nat 129:397–402

Scott DE (1994) The effect of larval density on adult demographic traits in *Ambystoma opacum*. Ecology 75:1383–1396

Seale DB (1987) Amphibia. In: Pandian TJ, Vernberg FJ (eds) Animal energetics, vol 2. Bivalvia through Reptilia. Academic Press, London, pp 467–552

Segura ET, Bandsholm UC, Bronstein A (1982) Role of the CNS in the control of the water economy of the toad *Bufo arenarum* Hensel. II. Adrenergic control of water uptake across the skin. J Comp Physiol [B] 146:101–106

Segura ET, Reboreda JC, Skorka A, Cuello ME, Petriella S (1984) Role of the CNS in the control of the water economy of the toad *Bufo arenarum* Hensel. III. Skin permeability increases to raised osmotic pressure of external milieau. J Comp Physiol [B] 154:573–578

Seibert EA, Lillywhite HB, Wassersug RJ (1974) Cranial coossification in frogs: relationship to rate of evaporative water loss. Physiol Zool 47:261–265

Semlitsch RD (1981) Terrestrial activity and summer home range of the mole salamander (*Ambystoma talpoideum*). Can J Zool 59:315–322

Semlitsch RD (1983a) Burrowing ability and behavior of salamanders of the genus *Ambystoma*. Can J Zool 61:616–620

Semlitsch RD (1983b) Structure and dynamics of two breeding populations of the eastern tiger salamander. Copeia 1983:608–616

Semlitsch RD (1985a) Analysis of climatic factors influencing migrations of the salamander *Ambystoma talpoideum*. Copeia 1985:477–489

Semlitsch RD (1985b) Reproductive strategy of a facutatively paedomorphic salamander *Ambystoma talpoideum*. Oecologia 65:305–313

Semlitsch RD (1987a) Relationship of pond drying to the reproductive success of the salamander *Ambystoma talpoideum*. Copeia 1987:61–69

Semlitsch RD (1987b) Density-dependent growth and fecundity in the paedomorphic salamander *Ambystoma talpoideum*. Ecology 68:1003–1008

Semlitsch RD (1987c) Paedomorphosis in *Ambystoma talpoideum*: effects of density, food, and pond drying. Ecology 68:994–1002

Semlitsch RD, Gibbons JW (1990) Effect of egg size on success of larval salamanders in complex aquatic environments. Ecology 71:1789–1795

Semlitsch RD, Pechmann JHK (1985) Diel pattern of migratory activity for several species of pond-breeding salamanders. Copeia 1985:86–91

Semlitsch RD, Wilbur HM (1988) Effects of pond drying time on metamorphosis and survival in the salamander *Ambystoma talpoideum*. Copeia 1988:978–983

Semlitsch RD, Wilbur HM (1989) Artificial selection for peadomorphosis in the salamander *Ambystoma talpoideum*. Ecology 43:105–112

Semlitsch RD, Scott DE, Pechmann JHK (1988) Time and size at metamorphosis related to adult fitness in *Ambystoma talpoideum*. Ecology 69:184–192

Semlitsch RD, Scott DE, Pechmann JHK, Gibbons JW (1993) Phenotypic variation in the arrival time of breeding salamanders: individual repeatability and environmental influence. J Anim Ecol 62:334–340

Sever DM (1991) Comparative anatomy and phylogeny of the cloacae of salamanders (Amphibia: Caudata). I. Evolution at the family level. Herpetologica 47:165–193

Sever DM (1992) Comparative anatomy and phylogeny of the cloacae of salamanders (Amphibia: Caudata). VI. Ambystomatidae and Dicamptodontidae. J Morphol 212:305–322

Sever DM (1993) Spermathecal cytology of *Ambystoma opacum* (Amphibia: Ambystomatidae) and the phylogeny of sperm sorage organs in female salamanders. J Morphol 217:115–127

Sever DM (1994) Observations on regionalization of secretory activity in the spermathecae of salamanders and comments on phylogeny of sperm storage in female amphibians. Herpetologica 50:383–397

Sever DM, Krenz JD, Johnson KM, Rania LC (1995) Morphology and evolutionary implications of the annual cycle of secretion and sperm storage in spermathecae of the salamander *Ambystoma opacum* (Amphibia: Ambystomatidae). J Morphol 223:35–46

Sexton OJ, Bizer JR (1978) Life history pattern of *Ambystoma tigrinum* in montane Colorado. Am Midl Nat 99:101–118

Sexton OJ, Phillips C, Bramble JE (1990) The effects of temperature and precipitation on the breeding migration of the spotted salamander (*Ambystoma maculatum*). Copeia 1990:781–787

Seymour RS (1973a) Physiological correlates of forced activity and burrowing in the spadefoot toad, *Scaphiopus hammondii*. Copeia 1973:103–115

Seymour RS (1973b) Energy metabolism of dormant spadefoot toads (*Scaphiopus*). Copeia 1973:435–445

Seymour RS (1973c) Gas exchange in spadefoot toads beneath the ground. Copeia 1973:452–460

Sharon R (1995) The reproductive system of the female salamander *Salamandra salamandra infraimmaculata* and its adaptation to water availability in different habitats. MSC Thesis, Technion, Haifa

Sherman E (1980a) Ontogenetic change in thermal tolerance of the toad *Bufo woodhousii fowleri*. Comp Biochem Physiol [A] 65:227–230

Sherman E (1980b) Cardiovascular responses of the toad *Bufo marinus* to thermal stress and water deprivation. Comp Biochem Physiol [A] 66:643–650

Shield JW, Bentley PJ (1973) Respiration of some urodele and anuran Amphibia. I. In water, role of the skin and gills. Comp Biochem Physiol [A] 46:17–28

Shoemaker VH (1964) The effects of dehydration on electrolyte concentrations in a toad, *Bufo marinus*. Comp Biochem Physiol 13:261–271

Shoemaker VH (1987) Osmoregulation in amphibians. In: Dejours P, Bolis L, Taylor CR, Weibel ER (eds) Comparative physiology: life in water and land. Liviana Press, Padova, pp 109–120

Shoemaker VH (1988) Physiological ecology of amphibians in arid environments. J Arid Environ 14:145–153

Shoemaker VH, Bickler PE (1979) Kidney and bladder function in a uricotelic treefrog (*Phyllomedusa sauvagei*). J Comp Physiol [B] 133:211–218

Shoemaker VH, McClanahan LL (1973) Nitrogen excretion in the larvae of a land-nesting frog (*Leptodactylus bufonius*). Comp Biochem Physion [A] 44:1149–1156

Shoemaker VH, McClanahan LL (1975) Evaporative water loss, nitrogen excretion and osmoregulation in phyllomedusine frogs. J Comp Physiol 100:331–345

Shoemaker VH, McClanahan LL (1980) Nitrogen excretion and water balance in amphibians of Borneo. Copeia 1980:446–451

Shoemaker VH, McClanahan LL (1982) Enzymatic correlates and ontogeny of uricotelism in tree frogs of the genus *Phyllomedusa*. J Exp Zool 220:163–169

Shoemaker VH, Nagy KA (1977) Osmoregulation in amphibians and reptiles. Annu Rev Physiol 39:449–471

Shoemaker VH, Baker MA, Loveridge JP (1989) Effect of water balance on thermoregulation in waterproof frogs (*Chiromantis* and *Phyllomedusa*). Physiol Zool 62:133–146

Shoemaker VH, McClanahan L, Ruibal R (1969) Seasonal changes in body fluids in a field population of spadefoot toads. Copeia 1969:585–591

Shoemaker VH, Balding D, Ruibal R, McClanahan LL (1972) Uricotelism and low evaporative water loss in a South American frog. Sci 175:1018–1020

Shoemaker VH, McClanahan LL, Withers PC, Hillman SS, Drewes RC (1987) Thermoregulatory response to heat in the waterproof frogs *Phyllomedusa* and *Chiromantis*. Physiol Zool 60:365–372

Shoemaker VH, Hillman SS, Hillyard SD, Jackson DC, McClanahan LL, Withers PC, Wygoda ML (1992) Exchange of water, ions and respiratory gases in terrestrial amphibians. In: Feder ME, Burggren WW (eds) Environmental physiology of the amphibians. Chicago University Press, Chicago, pp 125–150

Shoop CR (1965) Orientation of *Ambystoma maculatum*: movements to and from breeding ponds. Science 149:558–559

Shoop CR (1974) Yearly variation in larval survival of *Ambystoma maculatum*. Ecology 55:440–444

Showler DA (1995) Amphibian observations in Yemen, March-May 1993. Br Herpetol Soc Bull 52:22–25

Shpun S, Hoffman J, Katz U (1992) Anuran Amphibia which are not acclimable to high salt, tolerate high plasma urea. Comp Biochem Physiol [A] 103:473–477

Shpun S, Hoffman J, Nevo E, Katz U (1993) Is the distribution of *Pelobates syriacus* related to its limited osmoregulatory capacity? Comp Biochem Physiol [A] 105:135–139

Siebold CTV (1858) Über das Receptaculum Seminis der Weiblichen Urodelen. Z Wiss Zool 9:460–483

Sievert LM (1991) Thermoregulatory behaviour in the toads *Bufo marinus* and *Bufo cognatus*. J Therm Biol 16:309–312

Sinsch U (1991) Reabsorption of water and electrolytes in the urinary bladder of intact frogs (genus *Rana*). Comp Biochem Physiol [A] 99:559–565

Sinsch U, Eblenkamp B (1994) Allozyme variation among *Rana balcanica, R. levantina* and *R. ridibunda* (Amphibia: Anura). Z Zool Syst Forsch 32:35–43

Sinsch U, Seine R, Sherif N (1992) Seasonal changes in the tolerance of osmotic stress in natterjack toads (*Bufo calamita*). Comp Biochem Physiol [A] 101:353–360

Sivak JG, Warburg MR (1980) Optical metamorphosis of the eye of *Salamandra salamanadra*. Can J Zool 58:2059–2064

Sivak JG, Warburg MR (1983) Changes in optical properties of the eye during metamorphosis of an anuran, *Pelobates syriacus*. J Comp Physiol 150:329–332

Smith HM (1986) Ovoviviparity: spurious for ectotherms. BioScience 36:292

Smith JJB (1968) Hearing in terrestrial urodeles: a vibration-sensitive mechanism in the ear. J Exp Biol 48:191–205

Smits AW (1984) Activity patterns and thermal biology of the toad *Bufo boreas halophilus*. Copeia 1984:689–696

Snart RS, Dalton T (1973) Response of toad bladder to prolactin. Comp Biochem Physiol [A] 45:307–311

Snyder GK, Hammerson GA (1993) Interrelationships between water economy and thermoregulation in the canyon tree-frog *Hyla arenicolor*. J Arid Environ 25:321–329

Socino M, Ferreri E (1965) Sodium and potassium metabolism in the newt (*Triturus cristatus carnifex* Laur.). V. Effects of the treatment with aldosterone. Biochim Biol Sper 4:161–165

Soria MO, Berman DM, Coviello A (1987) Comparative effects of angiotensin II on osmotic water permeability in the toad (*Bufo arenarum*). Comp Biochem Physiol [A] 86:147–150

Spencer B (1896) Amphibia. In: Reports of the Horn expedition to central Australia, Part II. Zoology. Melville Mullen and Slade, Melbourne, pp 152–175

Spencer B (1928) Wanderings in Wild Australia, vol 1. Macmillan, London

Spencer B, Gillen FJ (1912) Across Australia, vol 1. Macmillan, London

Spight TM (1967a) The water economy of salamanders: water uptake after dehydration. Comp Biochem Physiol 20:767–771

Spight TM (1967b) The water economy of salamanders: exchange of water with the soil. Biol Bull 132:126–132

Spight TM (1968) The water economy of salamanders: evaporative water loss. Physiol Zool 41:195–203

Spotila JR, O'Connor MP, Bakken GS (1992) Biophysics of heat and mass transfer. In: Feder ME, Burggren WW (eds) Environmental physiology of the amphibians. University of Chicago Press, Chicago, pp 59–80

Stackhouse HL (1966) Some aspects of pteridine biosynthesis in amphibians. Comp Biochem Physiol 17:219–236

Steinwasher K (1978) Interference and exploitative competition among tadpoles of *Rana utricularia*. Ecology 59:1039–1046

Stebbins RC (1954) Amphibia and reptiles of western North America. McGraw-Hill, New York, 528 pp

Stenhouse SL (1985) Migratory orientation and homing in *Ambystoma maculatum* and *Ambystoma opacum*. Copeia 1985:631–7

Stenhouse SL, Hairston NG, Cobey AE (1983) Predation and competition in *Ambystoma* larvae: field and laboratory experiments. J Herpetol 17:210–220

Stewart MM (1967) Amphibians of Malawi. State University of New York, Press, New York, 163 pp

Stewart SG (1927) The morphology of the frog's kidney. Anat Rec 36:259–269

Stiffler DF (1981) The effects of mesotocin on renal function in hypophysectomized *Ambystoma tigrinum* larvae. Gen Comp Endocrinol 45:49–55

Stiffler DF (1988) Cutaneous exchange of ions in lower vertebrates. Am Zool 28:1019–1029

Stiffler DF (1994) Developmental changes in amphibian electrolyte and acid-base transport across skin. Isr J Zool 40:507–518

Stiffler DF, Alvarado RH (1974) Renal function in response to reduced osmotic load in larval salamanders. Am J Physiol 226:1243–1249

Stiffler DF, Hawk CT, Fowler BC (1980) Renal excretion of urea in the salamander *Ambystoma tigrinum*. J Exp Zool 213:205–212

Stiffler DF, Atkins BJ, Burt LD, Roach SC (1982) Control of renal function in larval *Ambystoma tigrinum*. J Comp Physiol [B] 149:91–97

Stiffler DF, Roach SC, Pruett SJ (1984) A comparison of the responses of the amphibian kidney to mesotocin, isotocin, and oxytocin. Physiol Zool 57:63–69

Stiffler DF, De Ruyter ML, Hanson PB, Marshall M (1986) Interrenal function in larval *Ambystoma tigrinum*. I. Responses to alterations in external electrolyte concentrations. Gen Comp Endocrinol 62:290–297

Stiffler DF, Ryan SL, Mushkot RA (1987) Interactions between acid-base balance and cutaneous ion transport in larval *Ambystoma tigrinum* (Amphibia: Caudata) in response to hypercapnia. J Exp Biol 130:389–404

Stinner JN, Shoemaker VH (1987) Cutaneous gas exchange and low evaporative water loss in the frogs *Phyllomedusa sauvagei* and *Chiromantis xerampelina*. J Comp Physiol [B] 157:423–427

Strubing H (1954) Über Vorzugstemperaturen von Amphibien. Z Morphol Ökol Tiere 43:357–386

Stuart LC (1951) The distributional implications of temperature tolerances and hemoglobin values in the toads *Bufo marinus* (Linnaeus) and *Bufo bocourti* Brocchi. Copeia 1951:220–229

Sullivan BK (1985) Sexual selection and mating system variation in anuran amphibians of the Arizona-Sonoran desert. Great Basin Nat 45:688–696

Sullivan BK (1989) Desert environments and the structure of anuran mating systems. J Arid Environ 17:175–183

Sweet G (1907) The anatomy of some Australian Amphibia. Part I. A. The openings of the nephrostomes from the coelom. B. The connections of the vasa efferentia with the kidney. Proc R Soc Victoria (NS) 20:222–249

Szarski H (1964) The structure of respiratory organs in relation to body size in Amphibia. Evolution 18:118–126

Taigen TL, Pough FH, Stewart MM (1984) Water balance of terrestrial anuran (*Eleutherdactylus coqui*) eggs: importance of parental care. Ecology 65:248–255

Tanner VM (1939) A study of the genus *Scaphiopus*. Great Basin Nat 1:3–26

Tarkhnishvili DN (1987) Growth dynamics in larvae of two species of Caucasus newts. Sov J Ecol 18:17–22

Taylor DH (1972) Extra-optic photoreception and compass orientation in larval and adult salamanders (*Ambystoma tigrinum*). Anim Behav 20:233–236

Taylor DH, Adler K (1978) The pineal body: site of extraocular perception of celestial cues for orientation in the tiger salamander (*Ambystoma tigrinum*). J Comp Physiol [A] 124:357–361

Taylor J (1983) Orientation of flight behavior of a neotenic salamander (*Ambystoma gracile*) in Oregon. Am Midl Nat 109: 40–49

Taylor P, Scroop GC, Tyler MJ, Davies M (1982) An ontogenetic and interspecific study of the renin-angiotensin system in Australian anuran Amphibia. Comp Biochem Physiol [A] 73:187–191

Tejedo M, Reques R (1994) Plasticity in metamorphic traits of natterjack tadpoles: the interactive effects of density and pond duration. Oikos 71:295–304

Tempel P, Himstedt W, Steinke S (1982) Spectral sensitivity of visual neurons and of phototactic behaviour in *Salamandra*. Zool Jahrb Physiol 86:401–412

Tester JR, Breckenridge WJ (1964) Population dynamics of the Manitoba toad, *Bufo hemiophrys*, in northwestern Minnesota. Ecology 45:592–601

Tevis L (1966) Unsuccessful breeding by desert toads (*Bufo punctatus*) at the limit of their ecological tolerance. Ecology 47:765–775

Thomas EO, Licht P (1993) Testicular and androgen dependence of skin gland morphology in the anurans, *Xenopus laevis* and *Rana pipiens*. J Morphol 215:195–200

Thomas EO, Tsang L, Licht P (1993) Comparative histochemistry of the sexually dimorphic skin glands of anuran amphibians. Copeia 1993:133–143

Thorson TB (1955) The relationship of water economy to terrestrialism in amphibians. Ecology 36:100–116

Thorson TB (1956) Adjustment of water loss in response to desiccation in amphibians. Copeia 1956:230–237

Thorson T, Svihla A (1943) Correlation of the habitats of amphibians with their ability to survive the loss of body water. Ecology 24:374–381

Toews D, Boutilier R, Todd L, Fuller N (1978) Carbonic anhydrase in the Amphibia. Comp Biochem Physiol [A] 59:211–213

Toledo RC, Jared C (1993a) The calcified dermal layer in anurans. Comp Biochem Physiol [A] 104:443–448

Toledo RC, Jared C (1993b) Cutaneous adaptations to water balance in amphibians. Comp Biochem Physiol [A] 105:593–608

Townsend DS, Stewart MM (1994) Reproductive ecology of the Puerto Rican frog *Eleutherodactylus coqui*. J Herpetol 28: 34–40

Toyoda F, Kikuyama S (1990a) Gonadotropin- and prolactin-treated newts prefer the water in which animals of the opposite sex have been kept. Zool Sci 7:1177

Toyoda F, Kikuyama S (1990b) Hormonal control of courtship behavior and sexual attractant secretion in the newt, *Cynops pyrrhogaster*. Zool Sci 7:1139

Toyoda F, Kikuyama S (1995) Hormonal induction of male-like courtship behavior in the female newt, *Cynops pyrrhogaster*. Zool Sci 12:815–818

Toyoda F, Matsuda K, Yamamoto K, Kikuyama S (1994) Involvement of endogenous prolactin in the expression of courtship behavior in the newt, *Cynops pyrrhogaster*. Gen Comp Endocrinol 102:191–196

Tracy CR, Dole JW (1969a) Orientation of displaced California toads, *Bufo boreas*, to their breeding sites. Copeia 1969:693–700

Tracy CR, Dole JW (1969b) Evidence of celestial orientation by California toads (*Bufo boreas*) during breeding migration. Bull South Calif Acad Sci 68:10–18

Tran DY, von Hoff KS, Hillyard SD (1992) Effects of angiotensin II and bladder condition on hydration behavior and water uptake in the toad, *Bufo woodhousei*. Comp Biochem Physiol [A] 103:127–130

Travis J (1984) Anuran size at metamorphosis: experimental test of a model based on intraspecific competition. Ecology 65:1155–1160

Trowbridge MS (1941a) Studies on the normal development of *Scaphiopus bombifrons* Cope. I. The cleavage period. Trans Am Microsc Soc 60:508–525

Trowbridge MS (1941b) Studies on the normal development of *Scaphiopus bombifrons* Cope. II. The later embryonic and larval periods. Trans Am Microsc Soc 60:66–83

Tufts BL, Toews DP (1986) Renal function and acid-base balance in the toad *Bufo marinus* during short-term dehydration. Can J Zool 64:1054–1057

Turner FH (1959) Some features of the ecology of *Bufo puctatus* in Death Valley, California. Ecology 40:175–181

Twitty VC (1959) Migration and speciation in newts. Science 130:1735–1743

Tyler MJ (1985) Reproductive modes in Australian Amphibia. In: Grigg G, Shine R, Ehmann H (eds) Biology of Australasian frogs and reptiles. Surrey Beatty, Sydney, pp 265–267

Tyler MJ (1994) Frogs of western New South Wales. In: Lunney D, Hand S, Reed P, Butcher D (eds) Future of the fauna of western New South Wales. Roy Zool Soc NSW, Mosman NSW, pp 155–160

Tyler MJ, Davies M (1986) Frogs of the Northern Territory. Government Printer Northern Territory, Alice Springs, 77 pp

Tyler MJ, Martin AA (1977) Taxonomic studies of some Australian leptodactylid frogs of the genus *Cyclorana* Steindachner. Rec S Aust Mus 17:261–276

Tyler MJ, Roberts JD, Davies M (1980) Field observations on *Arenophryne rotunda* Tyler, a leptodactylid frog of coastal sandhills. Aust Wildl Res 7:295–304

Tyler MJ, Watson GF, Martin AA (1981) The Amphibia: diversity and distribution. In: Keast A (ed) Ecological biogeography of Australia. Junk, The Hague, pp 1277–1301

Tyler MJ, Davies M, Martin AA (1982) Biology, morphology and distribution of the Australian fossorial frog *Cyclorana cryptotis* (Anura: Hylidae). Copeia 1982:260–264

Tyler MJ, Crook GA, Davies M (1983) Reproductive biology of the frogs of the Magela Creek system Northern Territory. Rec S Aust Mus 18:415–440

Uchiyama M (1980) Hypocalcemic factor in the ultimobranchial gland of the newt, *Cynops pyrrhogaster*. Comp Biochem Physiol [A] 66:331–334

Uchiyama M, Pang PKT (1981) Endocrine influence on hypercalcemic regulation in bullfrog tadpoles. Gen Comp Endocrinol 44:428–435

Ultsch GR (1976) Ecophysiological studies of some metabolic and respiratory adaptations of sirenid salamanders. In: Hughes GM (ed) Respiration of amphibious vertebrates, Academic Press, New York, pp 287–312

Uranga J (1973) Effect of glomerulopressin, oxytocin, and norepinephrine on glomerular pressure in the toad. Gen Comp Endocrinol 20:515–521

Uranga J, Sawyer WH (1960) Renal responses of the bullfrog to oxytocin, arginine vasotocin and frog neurohypophysial extract. Am J Physiol 198:1287–1290

Ushakov BP, Pashkova IM (1986a) Populational analysis of thermal responses. I. Changes in the heat resistance of muscle tissues and contractile muscle models of *Salamandra salamandra* larvae. J Therm Biol 11:167–173

Ushakov BP, Pashkova IM (1986a) Populational analysis of thermal responses. II. Changes in the rate of development of *Salamandra salamandra* larvae. J Therm Biol 11:175–180

Van Berkum F, Pough FH, Stewart MM, Brussard PF (1982) Altitudinal and interspecific differences in the rehydration abilities of Puerto Rican frogs (*Eleutherodactylus*). Physiol Zool 55:130–136

Van Beurden EK (1979) Gamete development in relation to season, moisture, energy reserve, and size in the Australian water-holding frog, *Cyclorana platycephalus*. Herpetologica 35:370–374

Van Beurden EK (1980) Energy metabolism of dormant Australian water-holding frogs (*Cyclorana platycephalus*). Copeia 1980:787–799

Van Beurden EK (1982) Desert adaptations of *Cyclorana platycephalus*: an holistic approach to desert-adaptation in frogs: In: Barker WR, Greenslade PJM (eds) Evolution of the flora and fauna of arid Australia. Peacock, pp 235–239

Van Beurden E (1984) Survival strategies of the Australian water-holding frog, *Cyclorana platycephalus*. In: Cogger HG, Cameron EE (eds) Arid Australia. Australian Museum, Sydney, pp 223–225

Van Dijk DE (1971) Anuran ecology in relation particularly to oviposition and development out of water. Zool Afr 6:119–132

Van Dijk DE (1972) The behaviour of Southern African anuran tadpoles with particular reference to their ecology and related external morphological features. Zool Afr 7:49–55

Van Dijk DE (1982) Anuran distribution, rainfall and soils in Southern Africa. South Afr J Sci 78:401–406

Vawda A (1978) The loss of sodium across the skin of the dehydrated toad, *Bufo regularis* (Reuss). J Herpetol Assoc Afr 17:4–9

Vellano C, Lodi G, Bani G, Sacerdote M, Mazzi V (1970) Analysis of the integumentary effect of prolactin in the hypophysectomized crested newt. Monit Zool Ital (NS) 4:115–146

Verhaagh M, Greven H (1982) Localization of calcium in fibriocytes associated with the "substantia amorpha" in the skin of the toad *Bufo bufo* (L.) (Amphibia, Anura). Acta Histochem (Jena) 70:139–149

Visser J, Cei JM, Gutierrez LS (1982) The histology of dermal glands of mating *Breviceps* with comments on their possible functional value in microhylids (Amphibia: Anura). S Afr J Zool 17:24–27

Volpe EP (1957) Embryonic temperature tolerance and rate of development in *Bufo valliceps*. Physiol Zool 30:164–176

Voûte CL, Meier W (1978) The mitochondria-rich cell of frog skin as hormone-sensitive "shunt-path". J Membr Biol 40:151–165

Voûte CL, Hanni S, Ammann E (1972) Aldosterone induced morphological changes in Amphibia epithelia. J Steroid Biochem 3:161–165

Voûte CL, Thummel J, Brenner M (1975) Aldosterone effect in the epithelium of the frog skin: a new story about an old enzyme. J Steroid Biochem 6:1175–1179

Wager VA (1986) Frogs of South Africa. Delta Books, Craighall, South Africa

Wake MH (1982) Diversity within a framework of constraints. Amphibian reproductive modes. In: Mossakowski D, Roth C (eds) Environmental adaptation and evolution. Fischer, New York, pp 87–106

Wake MH (1993) Evolution of oviductal gestation in amphibians. J Exp Zool 266:394–413

Waldman B (1981) Sibling recognition in toad tadpoles: the role of experience. Z Tierpsychol 56:341–358

Waldman B (1982) Sibling association among schooling toad tadpoles: field evidence and implication. Anim Behav 30:700–713

Waldman B (1985) Olfactory basis of kin recognition in toad tadpoles. J Comp Physiol [A] 156:565–577

Waldman B (1991) Kin recognition in amphibians. In: Hepper PG (ed) Kin recognition. Cambridge University Press, Cambridge, pp 162–219

Waldman B, Adler K (1979) Toad tadpoles associated preferentially with siblings. Nature 282:611–613

Walker RF, Whitford WG (1970) Soil water absorption capabilities in selected species of anurans. Herpetologica 26:411–418

Walls SC (1991) Ontogenetic shifts in the recognition of siblings and neighbours by juvenile salamanders. Anim Behav 42:423–434

Walls SC, Altig R (1986) Female reproductive biology and larval life history of *Ambystoma* salamanders: a comparison of egg size, hatchling size, and larval growth. Herpetologica 42:334–345

Walls SC, Blaustein AR (1994) Does kinship influence density dependance in a larval salamander? Oikos 71:459–468

Walls SC, Roudebush RE (1991) Reduced aggression toward siblings as evidence of kin recognition in cannibalistic salamanders. Am Nat 138:1027–1038

Walls SC, Belanger SS, Blaustein AR (1993a) Morphological variation in a larval salamander: dietary induction of plasticity in head shape. Oecologia 96:162–168

Walls SC, Beatty JJ, Tissot BN, Hokit DG, Blaustein AR (1993) Morphological variation and cannibalism in a larval salamander (*Ambystoma macrodactylum columbianum*). Can J Zool 71:1543–1551

Walls SC, Conrad CS, Murillo ML, Blaustein AR (1996) Agonistic behaviour in larvae of the northwestern salamander (*Ambystoma gracile*): the effects of kinship, familiarity and population source. Behaviour 133:1–20

Warburg MR (1965) Studies on the water economy of some Australian frogs. Aust J Zool 13:317–330

Warburg MR (1967) On thermal and water balance of three central Australian frogs. Comp Biochem Physiol 20:27–43

Warburg MR (1971a) On the water economy of Israel amphibians: the anurans. Comp Biochem Physiol [A] 40:911–924

Warburg MR (1971b) The water economy of Israel amphibians: the urodeles *Triturus vittatus* (Jenyns) and *Salamandra salamandra* (L.). Comp Biochem Physiol [A] 40:1055–1063

Warburg MR (1972) Water economy and thermal balance of Israeli and Australian Amphibia from xeric habitats. Symp Zool Soc Lond 31:79–111

Warburg MR (1986a) A relic population of *Salamandra salamandra* on Mt. Carmel: a ten-year study. Herpetologica 42:174–179

Warburg MR (1986b) Observation on a relic population of *Salamandra salamandra* on Mt. Carmel during eleven years. In: Roček Z (ed) Studies in herpetology. Charles University, Prague, pp 389–394

Warburg MR (1988) Adaptation of amphibians to life in xeric habitats. In: Ghosh PK, Prakash I (eds) Ecophysiology of desert vertebrates. Scientific Publishers, Jodhpur, pp 59–98

Warburg MR (1992a) Longevity of *Salamandra salamandra* on Mt. Carmel. In: Korsós Z, Kiss I (eds) Proc Sixth Ord Gen Meet S E H Budapest, pp 485–487

Warburg MR (1992b) Breeding patterns in a fringe population of fire salamanders, *Salamandra salamandra*. Herpetol J 2:54–58

Warburg MR (1994) Population ecology, breeding activity, longevity and reproductive strategies of *Salamandra salamandra* (Urodela, Salamandridae) during an 18-year long study of an isolated population on Mt. Carmel. Mertensiella Monogr 4:399–421

Warburg MR (1995) Hormonal effect on the osmotic, electrolyte and nitrogen balance in terrestrial Amphibia. Zool Sci 12:1–11

Warburg MR (1996) Pond fidelity in *Salamandra s. infraimmaculata* in a xeric environment. In: 6th Int Conf Preservation of our world in the wake of change, VI, B, 617–8 Jerusalem (in press)

Warburg MR, Degani G (1979) Evaporative water loss and uptake in juvenile and adult *Salamandra salamandra* (L.) (Amphibia: Urodela). Comp Biochem Physiol [A] 62:1071–1075

Warburg MR, Goldenberg S (1978a) Effect of oxytocin and vasotocin on water balance in two urodeles followed throughout their life cycle. Comp Biochem Physiol [A] 60:113–116

Warburg MR, Goldenberg S (1978b) The changes in osmoregulatory effects of prolactin during the life cycle of two urodeles. Comp Biochem Physiol [A] 61:321–324

Warburg MR, Lewinson D (1977) Ultrasturcture of epidermis of *Salamandra salamandra* followed throughout ontogenesis. Cell Tissue Res 181:369–393

Warburg MR, Rosenberg M (1990) Ion and water balance and their endocrine control in the aquatic amphibians. Fortschr Zool 38:385–403

Warburg MR, Degani G, Warburg I (1979) Growth and population structure of *Salamandra salamandra* (L.) larvae in different limnological conditions. Hydrobiologia 64:147–155

Warburg MR, Degani G, Warburg I (1978/1979) Ovoviviparity in *Salamandra salamandra* (L.) (Amphibia, Urodela) from northern Israel. Vie Milieu 28/29:247–257

Warburg MR, Lewinson D, Rosenberg M (1994a) Ontogenesis of amphibian epidermis. In: Heatwole H (ed) Amphibian biology vol I. The integument. Surrey Beatty, Sydney, pp 33–63

Warburg MR, Lewinson D, Rosenberg M (1994b) Structure and function of *Salamandra* skin and gills. Mertensiella Monogr 4:423–452

Wassersug RJ (1973) Aspects of social behavior in anuran larvae. In: Vial JL (ed) Evolutionary biology of the anurans. University of Missouri Press, Columbia, pp 274–298

Wassersug RJ, Lum AM, Potel MJ (1981) An analysis of school structure for tadpoles (Anura: Amphibia). Behav Ecol Sociobiol 9:15–22

Webb RG (1969) Survival adaptations of tiger salamanders (*Ambystoma tigrinum*) in the Chihuahuan Desert. In: Hoff CC Riedsel ML (eds) Physiological systems in semiarid environments. University of New Mexico Press, Albuquerque, pp 143–147

Weintraub JD (1974) Movement patterns of the red-spotted toad, *Bufo punctatus*. Herpetologica 30:212–215

Weintraub JD (1980) Selection of daytime retreats by recently metamorphosed *Scaphiopus multiplicatus*. J Herpetol 14:83–84

Wells KD (1977) The social behaviour of anuran amphibians. Anim Behav 25:666–693

Wentzell LA, McNeil SA, Toews DP (1993) The role of the lymphatic system in water balance processes in the toad *Bufo marinus* (L.). Physiol Zool 66:307–321

Werner YL (1988) Four modes of lightening of coloration in desert populations of the green toad, *Bufo viridis*. Zool Middle East 2:68–71

Whiteman HH (1994) Evolution of facultative paedomorphosis in salamanders. Q Rev Biol 69:205–221

Whitford WG (1969) Heart rate and changes in body fluids in aestivating toads from xeric habitats. In: Hoff CC, Riedsel ML (eds) Physiological systems in semiarid environments. University of New Mexico Press, Albuquerque, pp 125–133

Whitford WG (1973) The effects of temperature on respiration in the Amphibia. Am Zool 13:505–512

Whitford WG, Hutchison VH (1963) Cutaneous and pulmonary gas exchange in the spotted salamander, *Ambystoma maculatum*. Biol Bull 124:344–354

Whitford WG, Hutchison VH (1965a) Gas exchange in salamanders. Physiol Zool 38:228–242

Whitford WG, Hutchison VH (1965b) Effect of photoperiod on pulmonary and cutaneous respiration in the spotted salamander, *Ambystoma maculatum*. Copeia 1965:53–58

Whitford WG, Hutchison VH (1966) Cutaneous and pulmonary gas exchange in ambystomatid salamanders. Copeia 1966:573–577

Whitford WG, Hutchison VH (1967) Body size and metabolic rate in salamanders. Physiol Zool 40:127–133

Whitford WG, Massey M (1970) Responses of a population of *Ambystoma tigrinum* to thermal and oxygen gradients. Herpetologica 26:372–376

Whitford WG, Meltzer KH (1976) Changes in O_2 consumption, body water and lipid in burrowed desert juvenile anurans. Herpetologica 32:23–25

Whitford WG, Sherman RE (1968) Aerial and aquatic respiration in axolotl and transformed *Ambystoma tigrinum*. Copeia 1968:233–237

Wilbur HM (1977) Density-dependent aspects of growth and metamorphosis in *Bufo americanus*. Ecology 58:196–200

Wilbur HM (1980) Complex life cycles. Annu Rev Ecol Syst 11:67–93

Wilbur HM, Alford RA (1985) Priority effect in experimental pond communities: responses of *Hyla* to *Bufo* and *Rana*. Ecology 66:1106–1114

Wilbur HM, Collins JP (1973) Ecological aspects of amphibian metamorphosis. Science 182:1305–1314

Wilbur HM, Morin PJ, Harris RN (1983) Salamander predation and the structure of experimental communities: anuran responses. Ecology 64:1423–1429

Williams TA, Larsen JH (1986) New function for the granular skin glands of the eastern long-toed salamander, *Ambystoma macrodactylum columbianum*. J Exp Zool 239:329–333

Williamson I, Bull CM (1989) Life history variation in a population of the Australian frog *Ranidella signifera*: egg size and early development. Copeia 1989:349–356

Williamson I, Bull CM (1992) Life history variation in a population of the Australian frog *Ranidella signifera*: time of egg laying. J Herpetol 26:322–327

Williamson I, Bull CM (1995) Life history variation in a population of the Australian frog *Ranidella signifera*: seasonal changes in clutch parameters. Copeia 1995:105–113

Winokur RM, Hillyard S (1992) Pelvic cutaneous musculature in toads of the genus *Bufo*. Copeia 1992:760–769

Winston RM (1955) Identification and ecology of the toad *Bufo regularis*. Copeia 1955:293–302

Withers PC (1993) Metabolic depression during aestivation in the Australian frogs, *Neobatrachus* and *Cyclorana*. Aust J Zool 41:467–473

Withers PC, Hillman SS, Drewes RC (1984) Evaporative water loss and skin lipids of anuran amphibians. J Exp Zool 232:11–17

Withers PC, Hillman SS, Simmons LA, Zygmunt AC (1988) Cardiovascular adjustments to enforced activity in the anuran amphibian, *Bufo marinus*. Comp Biochem Physiol [A] 89:45–49

Withers PC, Hillman SS, Drewes RC, Sokol OM (1982a) Water loss and nitrogen excretion in sharp-nosed reed frogs (*Hyperolius nasutus*: Anura, Hyperoliidae). J Exp Biol 97:335–343

Withers PC, Louw G, Nicholson S (1982b) Water loss, oxygen consumption and colour change in "waterproof" reed frogs (*Hyperolius*). S Afr J Sci 78:30–32

Wittouck PJ (1972) Intensification par la prolactine de l'absorption d'ion sodium au niveau des branchies isolées de larves d'*Ambystoma mexicanum*. Arch Int Physiol Biochim 80:825–827

Wittouck PJ (1974) Tolérance ionique chez les larves de *Salamandra salamandra* L. (Amphibien, Urodele) immergées dans l'eau distillée. Arch Int Physiol Biochim 82:721–731

Wittouck PJ (1975a) Influence de la composition saline du milieu sur la concentration ionique du sérum chez l'axolotl intact et hypophysectomisé, effet de la prolactine. Gen Comp Endocrinol 27:169–178

Wittouck PJ (1975b) Action de la prolactine sur les cellules à mucus épidermiques chez l'axolotl intact et hypophysectomisé. Gen Comp Endocrinol 27:254–261

Wood CM, Munger RS, Toews DP (1989) Ammonia, urea and H^+ distribution and the evolution of ureotelism in amphibians. J Exp Biol 144:215–233

Woodward BD (1982) Tadpole competition in a desert anuran community. Oecologia 54:96–100

Woodward BD (1983) Predator-prey interactions and breeding-pond use of temporary-pond species in a desert anuran community. Ecology 64:1549–1555

Woodward BD (1986) Paternal effects on juvenile growth in *Scaphiopus multiplicatus* (the New Mexico spadefoot toad). Am Nat 128:58–65

Woodward BD (1987a) Interactions between Woodhouse's toad tadpoles (*Bufo woodhousii*) of mixed sizes. Copeia 1987:380–386

Woodward BD (1987b) Clutch parameters and pond use in some Chihuahuan desert anurans. Southwest Nat 32:13–19

Woodward BD (1987c) Paternal effects on offspring traits in *Scaphiopus couchi* (Anura: Pelobatidae). Oecologia 73:626–629

Woodward BD, Mitchell SL (1991) The community ecology of desert anurans. In: Polis GA (ed) The ecology of desert communities. University of Arizona Press, Tucson, pp 223–248

Wright PA (1995) Nitrogen excretion: three end products, many physiological roles. J Exp Biol 198:273–281

Wunderer H (1910) Die Entwicklung der äußeren Körperform des Alpensalamanders (*Salamandra atra* Laur.). Zool Jahrb Abt Anat Ontog Tiere 29:367–414

Wygoda ML (1984) Low cutaneous evaporative water loss in arboreal frogs. Physiol Zool 57:329–337

Wygoda ML, Williams AA (1991) Body temperature in free-ranging green tree frogs (*Hyla cinerea*): a comparison with "typical" frogs. Herpetologica 47:328–335

Yamashita K, Iwasawa H (1989) Metamorphic change in the skin induced by thyroxine at low temperature in *Hynobius nigrescens* larvae. Zool Sci 6:1194

Yanev KP, Wake DB (1981) Genic differentiation in a relict desert salamander, *Batrachoseps campi*. Herpetologica 37:16–28

Yokota SD, Hillman SS (1984) Adrenergic control of the anuran cutaneous hydroosmotic response. Gen Comp Endocrinol 53:309–14

Yorio T, Bentley PJ (1977) Asymmetrical permeability of the integument of tree frogs (Hylidae). J Exp Biol 67:197–204

Zaccone G, Fasulo S, Cascio PL, Licata A (1986) Enzyme cytochemical and immunocytochemical studies of flask cells in the amphibian epidermis. Histochemistry 84:5–9

Zamachowski W (1977) The water economy in some European species of anuran amphibians during the annual cycle. I. Water content of the organism. Acta Biol Cracov Ser Zool 20:181–289

Zamorano B, Cortes A, Salibian A (1988) Ammonia and urea excretion in urine of larval *Caudiverbera caudiverbera* (L.) (Anura, Amphibia). Comp Biochem Physiol [A] 91:153–155

Zhao EM, Adler K (1993) Herpetology of China. Publ Soc Study Amph Rep, Dayton, Ohio

Zoeller RT, Moore FL (1982) Duration of androgen treatment modifies behavioral response to arginine vasotocin in *Taricha granulosa*. Horm Behav 16:23–30

Zoeller RT, Moore FL (1986) Correlation between immunoreactive vasotocin in optic tectum and seasonal changes in reproductive behaviors of male rough-skinned newts. Horm Behav 20:148–154

Zoeller RT, Moore FL (1988) Brain arginine vasotocin concentrations related to sexual behaviors and hydromineral balance in an amphibian. Horm Behav 22:66–75

Zweifel RG (1968) Reproductive biology of anurans of the arid Southwest, with emphasis on adaptation of embryos to temperature. Bull Am Mus Nat Hist 140:1–64

Subject Index

Springer
and the
environment

At Springer we firmly believe that an international science publisher has a special obligation to the environment, and our corporate policies consistently reflect this conviction.

We also expect our business partners – paper mills, printers, packaging manufacturers, etc. – to commit themselves to using materials and production processes that do not harm the environment. The paper in this book is made from low- or no-chlorine pulp and is acid free, in conformance with international standards for paper permanency.

 Springer

Printing: Druckhaus Beltz, Hemsbach
Binding: Buchbinderei Schäffer, Grünstadt